AF595012

Vividness, Consciousness, and Mental Imagery

Vividness, Consciousness, and Mental Imagery

Making the Missing Links across Disciplines and Methods

Editor

Amedeo D'Angiulli

MDPI • Basel • Beijing • Wuhan • Barcelona • Belgrade • Manchester • Tokyo • Cluj • Tianjin

Editor
Amedeo D'Angiulli
Carleton University
Canada

Editorial Office
MDPI
St. Alban-Anlage 66
4052 Basel, Switzerland

This is a reprint of articles from the Special Issue published online in the open access journal *Brain Sciences* (ISSN 2076-3425) (available at: https://www.mdpi.com/journal/brainsci/special_issues/Vividness_Consciousness_Imagery).

For citation purposes, cite each article independently as indicated on the article page online and as indicated below:

LastName, A.A.; LastName, B.B.; LastName, C.C. Article Title. *Journal Name* **Year**, *Volume Number*, Page Range.

ISBN 978-3-0365-0412-4 (Hbk)
ISBN 978-3-0365-0413-1 (PDF)

Contents

About the Editor

Amedeo D'Angiulli is Professor of Developmental Social and Cognitive Neuroscience. He is also Director of the Neuroscience of Imagination, Cognition and Emotion Research (NICER) Lab at Carleton University, Department of Neuroscience, Ottawa, Canada. Throughout his career, he has been the recipient of numerous awards (including the Killam Fellowship, Canada Research Chair (Tier II), and the IBRO-UNESCO Fellowship in Learning Sciences) and funding from National and Provincial agencies in Canada, as well as internationally. He has made over one hundred research contributions of different kinds, with over ninety peer-reviewed publications, and serves/ed in the boards of several journals as Associate, Consulting Editor and/or Reviewer, or Chief Editor (Imagination, Cognition and Personality, SAGE), as well as in boards of AI companies, schools and non-profit NGOs.

Editorial

Vividness, Consciousness and Mental Imagery: A Start on Connecting the Dots

Amedeo D'Angiulli

Department of Neuroscience, Carleton University, Ottawa, ON K1S 5B6, Canada; amedeo.dangiulli@carleton.ca; Tel.: +1-613-520-2600 (ext. 2954)

Received: 21 July 2020; Accepted: 30 July 2020; Published: 31 July 2020

Over twenty years ago, Baars [1] noted that, "The strategy of treating consciousness as a variable has now become standard in the study of subliminal vision, blindsight, and implicit cognition. We can easily apply it to mental imagery—yet we rarely do so". He concluded that, "As a result, even as consciousness research thrives in other domains, we have very little firm evidence about the conscious dimension of mental imagery". In many respects, Baar's conclusion still stands.

Today in many studies, mental images are still either treated as conscious by definition, or as empirical operations implicit to completing some type of task, such as the measurement of reaction time in mental rotation, where an underlying mental image is assumed, but there is no direct determination of whether it is conscious or not.

The vividness of mental images is a potentially helpful construct which may be suitable, as it may correspond to consciousness or aspects of the consciousness of images. There is currently a surge of interest in vividness in cognitive neuroscience and neuroimaging literature (see [2] for a review). It seems that a general implicit assumption is that vivid images are conscious, and it is possible that the least vivid images are effectively unconscious or that they become such once a threshold (e.g., the "no image" in the Vividness of Visual Imagery Questionnaire) is reached. Thus, it is still unclear whether the vividness dimension may in fact be a kind of "disguised correlate of consciousness" [1] or if instead it might be a supramodal metacognitive dimension not necessarily associated with imagery. However, even from studies using a vividness approach in neuroscience, the conscious dimension of mental imagery is not explicitly or fully tackled head on. Indeed, when a proper exhaustive literature search is conducted, it leads back to a paper by David Marks [3] for a glimpse of what a theoretical discussion of the missing links might look like.

In this context, a complicating factor seems to be the surprising variety in what is meant by the term vividness or how it is used or theorized. Some authors do not mention imagery or consciousness at all when using the term vividness, but associate it to various forms of memory such as prospective, episodic, autobiographical, or to aliased processes not literally called imagery but, for example, imagining or visualizations or simulations. Similarly, replacement constructs for vividness have been offered, for example, in terms of the strength of imagery or semantic long-term memory contents. In all these cases, it is not really clear what is achieved by replacing one label with another or replacing a research tradition with another. We are still left with the gaps pointed out by Baars.

To start filling some of the gaps, the goal of the present Special Issue was to create a forum where authors could fully explore through sound research the missing theoretical and empirical links between vividness, consciousness, and mental imagery across disciplines.

Craver-Lemley and Reeves [4] opened this Special Issue with the study of the effect of a sweetness blocker, *Gymnema sylvestre* (a known selective suppressor of sweetness only, but not of other taste sensations), on the synesthetic taste vividness of a rare color-gustatory synesthete, E.C., for whom specific colors elicit unique tastes. Given that E.C.'s concurrent color experience can be sweet, Gymnema sylvestre provides a unique opportunity to test the role of sensory modulation in this form of synesthesia. Blocking E.C.'s sweetness receptors while the tongue was otherwise unstimulated left other taste components of the synesthesia unaltered but initially reduced her synesthetic sweetness (vividness), which suggests a peripheral modulation of the synesthetic illusion. These authors contend that although their data are from a single, very rare subject, and are speculative, this should not detract from the theoretical importance of E.C.'s synesthesia, "in which—as is otherwise almost never the case—an all-or-nothing sensory manipulation could be applied to the concurrent". In light of this, the authors plea for the study of more cases of this sort, so that the role of an active bottom-up "tonic" signal can be further verified.

Remaining in the theme of "perceptual" vividness, Pinna and Conti [5] presented a novel take on amodal completion (AC), defined as vivid completion in a single continuous object of the visible parts of an occluded shape despite that portions of its boundary contours are not actually seen. According to their perspective, AC is a visual phenomenon and not a process (that is, not the final result of perceptual processes and grouping principles). To demonstrate this guiding proposition, the authors investigated the role of contrast polarity in newly devised stimuli assessed through the Gestalt experimental phenomenology approach. The results demonstrated the domination of the contrast polarity against good continuation, T-junctions, and regularity, even leading to the identification of a new type of junction (I-junctions). The authors defended the primacy of contrast polarity against traditional accounts of AC such as the Prägnanz principle, Helmholtz's likelihood, and Bayesian inference.

To appreciate the difference in perspectives between the experimental phenomenology approach adopted by Pinna and Conti and the computational contenders, in this Special Issue, the readers can find a very stimulating commentary debate between these authors and the (unfortunately) late Peter van der Helm. The latter author claimed [6] that simplicity and likelihood approaches can account for the phenomena described by Pinna and Conti; and argues that Pinna and Conti's approach erroneously confounds simplicity and likelihood. In their rebuttal, Pinna and Conti [7] rejected van der Helm's interpretation of their hypotheses and methodological approach, noting that none of van der Helm's comments were directly related to their stimuli. The authors clarified the complementarity of their approach with other potentially useful perspectives in explaining the yet unaccounted for stimuli they presented in the original work and their rebuttal commentary, including a possible important role of Bayesian simulations in contributing to the renewed debate on AC.

Going from perception to memory, Lefebvre and D'Angiulli [8] reported a study of the joint influences of familiarity and vividness on image generation from verbal stimuli, as characterized by image formation latency (RTs) and rate of incidental recall. Surprisingly, they found that two strong or weak codes are better than two medium ones, and that matching levels of strong or low vividness and familiarity result in faster imagery and more recalled stimuli than when both codes are of middling intensity. To explain these results, the authors proposed a dualistic neuropsychological model of memory consolidation that mimics the global activity in two large resting-state brain networks (RSNs), the default mode network (DMN) and the task-positive network (TPN). As discussed by the authors, these findings may have some applications to the clinical field for developing neurophenomenological markers of core memory deficits currently hypothesized to be shared across multiple psychopathological conditions.

Probably one of the most influential figures in vividness research, David Marks [9] took a renewed look at the systematization of the field spanning over at least four decades of work. He presented a general theory, synthesizing the reciprocal interconnections within a dynamic system involving imagery, affect, and action and consciousness as the central executive. The fundamental assumption of this general theory is that the primary motivation of all of consciousness and intentional behavior is psychological homeostasis. In this context, the degree of vividness of inner imagery is beneficial to imagining, remembering, thinking, predicting, planning, and acting. Marks reviewed key supporting work in cognitive neuroscience which directly validates introspective reports on vividness.

Thorudottir and Sigurdardottir and their team of colleagues [10] report the intriguing neuropsychological and neuroimaging case study of an architect (PL518) who lost his ability for visual imagery (aphantasia) following a bilateral posterior cerebral artery (PCA) stroke. When comparing the neuropsychological profile and structural magnetic resonance imaging (MRI) for PL518 to patients with either a comparable background (an architect) or bilateral PCA lesions, the authors found in all patients substantial shared lesions and cognitive deficits (except aphantasia, only occurring in PL518). The only selective lesions unique to PL518 were a small area in the left fusiform gyrus and part of the right lingual gyrus. These authors concluded that these areas might play a necessary role in the cerebral network involved in visual imagery.

The Special Issue concludes with an empirical examination of the vast literature. Haustein and colleagues [11] performed a bibliometric analysis of the peer-reviewed literature on vividness between 1900 and 2019 indexed by the Web of Science and compared it with the same analysis of publications on consciousness and mental imagery. While the citation patterns for papers in these three subjects are similar, the underlying topical concepts rarely overlap (co-occur) explicitly in the literature. The field of psychology dominates the topic of vividness, even though the total number of publications containing that term is small and the concept occurs in several other disciplines such as computer science and artificial intelligence. The authors suggest that without a coherent unitary framework for the use of vividness in research, important opportunities for advancing the field might be missed. The alternative is an evidence-based framework (such as bibliometric analytic methods) to guide transdisciplinary research on vividness and to resolve the challenge of conceptual, methodological, and terminological inconsistencies and differences amongst published research in multidisciplinary fields.

In conclusion, this Special Issue represented just a small cross-section of the multidimensional research landscape wherein vividness, consciousness, and mental imagery implicitly or sometimes explicitly co-exist. We are still very far from illuminating its very complex and dynamic nature rich of diverse methodological, empirical, and theoretical relationships. However, we believe we have at least planted the seeds to go a step further.

Conflicts of Interest: The author declares no conflict of interest.

References

1. Baars, B.J. When Are Images Conscious? The Curious Disconnection between Imagery and Consciousness in the Scientific Literature. *Conscious. Cogn.* **1996**, *5*, 261–264. [CrossRef] [PubMed]
2. Runge, M.; Cheung, M.W.-L.; D'Angiulli, A. Meta-analytic comparison of trial-versus questionnaire-based vividness reportability across behavioral, cognitive and neural measurements of imagery. *Neurosci. Conscious.* **2017**, *1*. [CrossRef] [PubMed]
3. Marks, D.F. Consciousness, mental imagery and action. *Br. J. Psychol.* **1999**, *90*, 567–585. [CrossRef]
4. Craver-Lemley, C.; Reeves, A. Taste Modulator Influences Rare Case of Color-Gustatory Synesthesia. *Brain Sci.* **2019**, *9*, 186. [CrossRef] [PubMed]
5. Pinna, B.; Conti, L. The Limiting Case of Amodal Completion: The Phenomenal Salience and the Role of Contrast Polarity. *Brain Sci.* **2019**, *9*, 149. [CrossRef] [PubMed]
6. Van der Helm, P.A. Dubious Claims about Simplicity and Likelihood: Comment on Pinna and Conti (2019). *Brain Sci.* **2020**, *10*, 50. [CrossRef] [PubMed]
7. Pinna, B.; Conti, L. On the Role of Contrast Polarity: In Response to van der Helm's Comments. *Brain Sci.* **2020**, *10*, 54. [CrossRef] [PubMed]
8. Lefebvre, E.; D'Angiulli, A. Imagery-Mediated Verbal Learning Depends on Vividness–Familiarity Interactions: The Possible Role of Dualistic Resting State Network Activity Interference. *Brain Sci.* **2019**, *9*, 143. [CrossRef] [PubMed]
9. Marks, D.F. I Am Conscious, Therefore, I Am: Imagery, Affect, Action, and a General Theory of Behavior. *Brain Sci.* **2019**, *9*, 107. [CrossRef] [PubMed]

10. Thorudottir, S.; Sigurdardottir, H.M.; Rice, G.E.; Kerry, S.J.; Robotham, R.J.; Leff, A.P.; Starrfelt, R. The Architect Who Lost the Ability to Imagine: The Cerebral Basis of Visual Imagery. *Brain Sci.* **2020**, *10*, 59. [CrossRef] [PubMed]
11. Haustein, S.; Vellino, A.; D'Angiulli, A. Insights from a Bibliometric Analysis of Vividness and Its Links with Consciousness and Mental Imagery. *Brain Sci.* **2020**, *10*, 41. [CrossRef] [PubMed]

Article

Taste Modulator Influences Rare Case of Color-Gustatory Synesthesia

Catherine Craver-Lemley [1],* and Adam Reeves [2]

[1] Department of Psychology, Elizabethtown College, Elizabethtown, PA 17022, USA
[2] Department of Psychology, Northeastern University, Boston, MA 02115, USA
* Correspondence: lemleyce@etown.edu

Received: 3 July 2019; Accepted: 30 July 2019; Published: 31 July 2019

Abstract: We investigated the effect of a sweetness blocker on the synesthetic taste experience of a rare color-gustatory synesthete, E.C., for whom specific colors elicit unique tastes. Blocking E.C.'s sweetness receptors while the tongue was otherwise unstimulated left other taste components of the synesthesia unaltered but initially reduced her synesthetic sweetness, which suggests a peripheral modulation of the synesthetic illusion.

Keywords: color-gustatory synesthesia; taste; taste modulator; synesthesia

---

1. Introduction

Synesthesia occurs when the stimulation of one sensory modality (the "inducer") triggers an involuntary and simultaneous perception in the same or in another modality (the "concurrent") [1]. Synesthesia can involve cross-linkage among any of the sense modalities. For example, sound-color synesthetes may see colored shapes in response to hearing sounds such as music or voices, and grapheme-color synesthetes see unique colors when viewing written words, letters or numerals. Here we report on "E.C.", a highly unusual case of color-gustatory synesthesia, in which color acts as the inducing stimulus for concurrent gustatory sensations. In the case of the slightly less unusual lexical-gustatory synesthesia, written or spoken words elicit taste sensations [2–8]. As color is frequently the concurrent [9], but rarely the inducer [10], our case, in which color is the inducer, is also untypical in this respect. Indeed, we are aware of only two similar cases, "S" and "T.K.", reported a century apart [11,12]. S experienced both sound-color synesthesia and a reportedly less intense color-gustatory synesthesia [11]. S stated " ... when I put my mind intently on colors, I taste them. I can taste blue" (pp. 40–41). Little else was described regarding S's color-gustatory synesthesia in the original report. Fortunately, Nikolinakos et al. [12], provided detailed documentation of TK's color-gustatory experience, allowing for some comparison with our case.

Craver-Lemley and Mastrangelo [13], had already demonstrated some degree of cognitive modulation of E.C.'s synesthesia, as indicated by her experience in viewing the face/urn display in Figure 1. Viewing this figure, she reported that the blue "tasted very sweet" and the green tasted "fresh, like rain with no humidity, a hint of cilantro, slightly tangy". Both tastes were never experienced simultaneously; instead, the taste depended upon the color of the figure that was perceived—faces or urn. Her synesthetic tastes 'flipped' along with her visual reversals; she only experienced synesthesia for the color that forms the figure. In this respect, her form of synesthesia is generally consistent with lexical-gustatory synesthesia, which also shows cognitive influences [7,8]. For example, Bankieris and Simner [2], showed that particular phonological features of inducer words predict certain categories of taste in the concurrent, although the latter finding reflected sound symbolic patterns in English (onomatopoeia), whereas the former finding (Figure 1) primarily reflects the disposition of attention.

Figure 1. A color-gustatory synesthete, E.C., reported that the blue "tasted very sweet" and the green tasted "fresh, like rain with no humidity, a hint of cilantro, slightly tangy". In this figure, both tastes were not experienced simultaneously. The taste depended upon the color of the figure that was perceived—faces or urn. Her synesthetic tastes 'flipped' along with her visual reversals.

The purpose of the present study was to determine whether E.C.'s color-gustatory synesthesia could also be modulated at a sensory level. We investigated this by determining whether a sweetness blocker, *Gymnema sylvestre*, when applied to the tongue, could influence E.C.'s synesthetic sweetness. E.C. reported that her synesthesia occurs in her mouth and tongue, which is consistent with accounts from other individuals experiencing synesthetic taste [3,4,11,14] Clearly, this does not imply a sensory origin—the synesthetic taste originates as an association in the brain, even if it is referred back to the mouth. However, some form of sensory modulation might be possible. An apparently adequate test by applying a real taste to the tongue while simultaneously inducing a taste could be problematic because E.C. does not experience synesthetic taste blends, as illustrated by Figure 1, rather she experiences one taste or the other, but not a combination. Therefore, the hunt was on to find a taste modulator that could affect the tongue in the complete absence of gustatory stimuli (food or liquid). The sweetness blocker, *Gymnema sylvestre*, was used because it blocks only sweetness, leaving other tastes unaffected.

2. Method

2.1. Participant

E.C. is a right-handed 27-year-old female graduate student who experiences color-gustatory synesthesia. Whenever E.C. views a specific color, she automatically experiences distinct taste percepts, sometimes accompanied by texture and emotions (Tables 1 and 2). Her reports are consistent over years, as is generally typical of synesthetes. Sweet is her most frequent synesthetic taste, sour and bitter occur infrequently, having few inducers, and saltiness does not occur at all. Markedly, T.K. also reported an absence of saltiness with his synesthesia [12]. Additionally, E.C. often describes different colors as eliciting a 'spicy' taste. She also experiences tactile sensations (such as a "sand paper" or "grainy") in her mouth when viewing some colors. Metallic colors in Crayola® crayons will induce a "crackling" sensation that she likens to Pop Rocks® candies and finds to be very enjoyable. She reports that her synesthetic taste lasts as long as she views a color. E.C. reports that she was unaware that she had synesthesia until the topic was discussed in one of her classes. She is not aware of any of her relatives being synesthetes.

Table 1. Illustrative concurrents.

Inducer Color	Synesthete's Concurrent
Burnt orange	Unpleasant, tastes like bitter salad greens, peppery, too much pepper
Spring green	Fruity and minty
Blue violet	Honey and sugar taste
Brown	Thick cream, heavy whip cream

Table 2. Experimental ratings.

Inducer Color	Baseline Comments	Sweet Rating ^	Post-Sweet Blocker Comments	Sweet Rating ^
Yellow	Very sweet, like cheap grocery store cupcake frosting	3.75	Not as sweet but still unpleasant, flaky, gooey	0
Periwinkle	Like it! Sweet and spicy, maybe tastes like a gardenia would	3.25	Still a favorite	1.75
Shocking Pink	Sweet, but not as overwhelming as yellow, like a Lifesaver® candy—sorbet and mint, no texture, cheerful	3.0	No taste	0.5
Steel Blue	Sweet and spicy, bold, textural experience like Pop Rocks® candy (crackling and popping sensation), pleasant	3.25	Not pleasant, even though the Pop Rocks® texture is there, taste is turned down	2.0
Ruby	More spicy than sweet, tangy, Pop Rocks® experience, energizing	3.0	Decrease in the flavor, but texture is there	1.25

^ Mean ratings from the first two *Gymnema sylvestre* sessions.

This research was conducted in accordance with the American Psychological Association's standards for the ethical treatment of subjects and with the approval of the Institutional Review Board (IRB) for Human Research of Elizabethtown College, IRB #FA09-05. Before participating, E.C. was informed that she could leave the experiment at any time without penalty. She was informed of the full procedure of the experiment, including being asked to describe her synesthesia, and that she would be given health food store teas or powders. She signed the informed consent form approved by the IRB.

2.2. Materials

Crayola® crayons were used to induce E.C.'s synesthesia. The approximate R, G, B values provided by the manufacturer for a color monitor in 2009 were Yellow (252, 232, 131), Periwinkle (197, 208, 230), Shocking Pink (251, 126, 253), Metallic FX Steel Blue (0, 129, 171), Metallic FX Big Dip O' Ruby (156, 37, 66). The *Gymnema sylvestre* tea and a gunpowder green tea (*Camellia sinensis*, used for a placebo condition) were purchased from Starwest Botanicals, Inc. Both loose teas were brewed according to the manufacturer's instructions and presented at approximately room temperature. We also used a powdered form of *Gymnema sylvestre* purchased in 400 mg capsules from Swanson Health Products. In this case, E.C. spread the contents of two capsules over the surface of her tongue. *Gymnema sylvestre* is a known suppressor of sweetness, but not of other taste sensations, which acts to inhibit sucrose receptors on the tongue [15]. Given that E.C.'s concurrent can be sweet, *Gymnema sylvestre* provides a unique opportunity to test the role of sensory modulation in this form of synesthesia.

2.3. Procedure

Across six sessions, over a 16-month period, E.C. was presented with either the sweetness blocker, *Gymnema sylvestre* (tea that blocks sugars from activating sweet receptors on the tongue) or a placebo, *Camellia sinensis* (gunpowder green tea) that looks and tastes like the experimental tea. The blocker was given in four of the sessions and the control tea was given in the remaining two sessions. Immediately prior to each session, she was also asked to provide a report of the tastes elicited by the crayons after swishing distilled water around her mouth to provide a baseline. During the first, second, and fourth experimental session, E.C. was instructed to swish an ounce of the tea around the inside of her mouth for 30 s and then to swallow. On the third experimental session, *Gymnema sylvestre* powder made from ground tea leaves was placed directly on E.C.'s tongue instead of having her drink the tea. After she ingested the tea, she was individually presented with five different colors (see above) in random order. These colors were selected because pre-testing revealed that they typically elicit the sweetest synesthetic tastes for E.C., and because none of these colors evoke an aversive taste experience for her. Using crayons also allowed us to include stimuli that trigger the "crackling" sensation, along with taste. She reported that she was familiar with the tastes associated with the crayon colors yellow, periwinkle, and shocking pink, but that the crackling sensations and colors associated with the metallic ruby and steel blue crayons were novel.

In all conditions, after each color was presented, E.C. rated the intensity of her synesthetic sweetness on a scale from 0 to 5 (0 = none; 1 = weak to 5 = strong). She also described her synesthetic experience while viewing each color. Upon the conclusion of each session, E.C. was given chocolate candies to consume. After the *Gymnema sylvestre* sessions, she reported that the candy tasted bland, "really strange" and not sweet, verifying that the *Gymnema sylvestre* affected her "normal" taste sensation (as it does with non-synesthetes).

At the start of the fourth experimental session, we asked E.C. to memorize the codes for the five experimental colors until she could evoke a mental image of each color when she heard the associated code. We then asked her to generate a mental image of each color and to rate the sweetness of the concurrent induced by the mental image. After a 30-min break, she then participated in the same baseline versus *Gymnema sylvestre* comparison as was run in the previous three experimental sessions. The sessions took 40 min to an hour to run.

3. Results

E.C. initially provided descriptions and ratings for the experimental condition that were less sweet, indicating that her synesthetic sweetness was modified by the *Gymnema sylvestre* (Table 2). Her textural experience was not affected. E.C. appeared disconcerted that her colors were not eliciting her familiar tastes. E.C. described her experience as "like seeing a little girl that you know, opening her mouth to speak and hearing an old man's voice instead." The *Gymnema sylvestre* tea diminished or eliminated the synesthetic sweetness E.C. typically experiences when viewing these colors. The colors still triggered some taste and texture, however, indicating that the tea influenced the sweetness only of E.C.'s synesthetic experience.

Although the first two presentations of the *Gymnema sylvestre* tea impacted E.C.'s synesthesia, while the placebo did not, when we later placed *Gymnema sylvestre* powder directly on E.C.'s tongue (this occurred a month after the second presentation of the tea), it had no effect. We thought that perhaps the powder method of administering the *Gymnema sylvestre* might account for the difference, which may be less effective [15], so we tested her three weeks later with the *Gymnema sylvestre* tea, and again there was no effect. Critically, her descriptions of the tastes, not just her sweetness ratings, no longer differed between baseline and tea. Figure 2 shows this result by plotting her sweetness ratings in the baseline and blocker conditions across experimental sessions 1 through 4. A two-way ANOVA of the session by the condition (baseline or blocker) confirmed the visual impression given by Figure 2, in that the main effects of the session ($F_{3,32} = 36.1$, $p < 0.05$) and condition ($F_{1,32} = 31.5$, $p < 0.05$) were significant, as was their interaction ($F_{3,32} = 10.2$, $p < 0.05$). The significant interaction is accounted for by the elimination of the blocker effect in sessions 3 and 4.

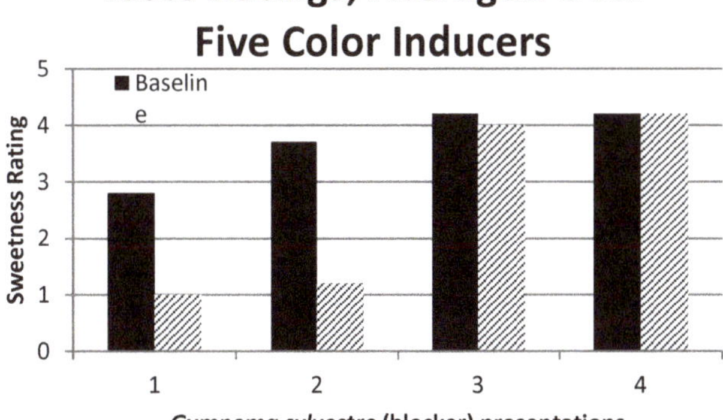

Figure 2. E. C.'s synesthetic sweetness ratings for the baseline and sweetness blocker conditions across four experimental sessions over a 16-month period. The blocker stopped influencing the synesthetic sweetness after the first two presentations.

We also found that E.C.'s mental images of the experimental colors yielded the corresponding synesthetic tastes. Previously, E.C. had reported that, while imagined colors could trigger her synesthesia, the mental imagery only elicited a "ghost" of her typical synesthetic taste. However, her synesthetic sweetness ratings due to imagery were somewhat higher than the baseline ratings she provided when actually viewing the crayons. A summary of these findings is provided in Figure 3, which plots the mean sweetness ratings averaged over the baseline conditions, in the first two and last two blocker sessions, and with imagery. This finding is in line with other reports provided by synesthetes, where the mental image of an inducer can sometimes trigger a more vivid synesthetic experience than the original inducer [16] (p. 198), but it is in contrast to Nikolinakos et al.'s [12] subject, T.K., whose mental images of colors did not produce his gustatory synesthesia.

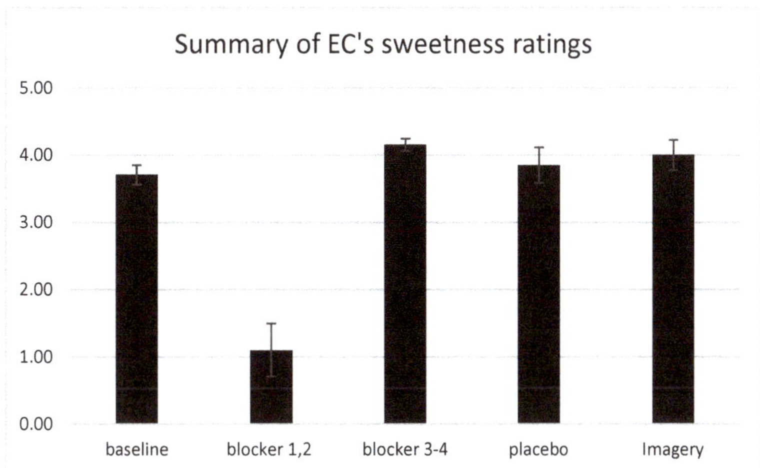

Figure 3. E.C. viewed five color displays and rated the synesthetic sweetness intensity after drinking water (baseline condition; average of six sessions), drinking a taste modulator which blocks sweetness (blocker condition; average of sessions 1–2 and 3–4), or drinking a placebo (average of two sessions), and after simply evoking mental images of the five colors (in experimental session 4). Bars show ±1 standard error of each mean.

4. Discussion

We found that the *Gymnema sylvestre* tea suppressed E.C.'s synesthetic sweetness only at the time of the first and second presentations. One way of explaining the reduction in the intensity of the concurrent after repeated administration of the sweetness blocker is in terms of E.C.'s expectations, or desire to please the experimenter [17]. However, E.C. was given the green tea in the very first of her sessions, and the tea had no effect. The blocker was given in subsequent sessions, and, as mentioned, it tasted like the green tea. Therefore, the strong reduction in the synesthetic taste in the first experimental session (see Figure 2) is unlikely to be an expectancy effect.

We therefore consider possible physiological effects. One is that spontaneous neural activity generated by the tongue in the absence of stimulation (a tonic 'bottom-up' signal) is needed to activate the gustatory cortex sufficiently for synesthetic taste sensations to occur. That is, sweetness cells in the taste buds must be spontaneously active for the color impression to be able to generate a concurrent taste through a synesthetic process. Spontaneous activity is known to occur; for example, 82% of taste receptors (fungiform papillae) on the tongue of the rat are spontaneously active [18]. The hypothesis that such spontaneous activity supports gustatory synesthesia is novel, and not based on known physiology. If true, a blocker such as *Gymnema sylvestre*, which decreases spontaneous activity in sweetness cells, would attenuate the spontaneous bottom-up signal and thereby eliminate the concurrent taste. An additional possibility is that the blocker may have affected the salivary response of the tongue and palette that is normally associated with sweetness, indirectly modulating the synesthetic sweetness. Other explanations may be possible. However, any explanation must also account for why E.C.'s other synesthetic tastes (bitter and sour) were not affected. The colors associated with those tastes continued to have their normal effects throughout testing.

If any such physiological account is correct, why did repetition in experimental sessions 3 and 4 restore the concurrent? The direct effect of the blocker on the sweetness of the candy bar that E.C. ate at the end of each session did not adapt out—the blocker was effective at the start of the experiment and at the end. It therefore seems unlikely that the effect of the blocker on the synesthetic response adapted out. A hypothesis we favor is that the long-established color taste associations in memory, as revealed in the imagery test in the last session, eventually over-ruled the synesthetic effect of the blocker. If so, E.C.'s synesthesia may be unique in demonstrating a change in the control of synesthesia over repeated administrations from more sensory control to more cognitive control, a change not previously mentioned in the literature possibly because of the difficulty in eliminating the sensory basis of the concurrent, a difficulty removed for E.C. by the fortuitous availability of *Gymnema sylvestre*.

A basic issue in the study of synesthesia concerns the relationship between the concurrent, which is by definition imaginary, and everyday mental images. Craver-Lemley and Reeves [16], argued that they are distinct, as concurrents are obligatory, experienced only in the presence of the inducer, and of uncontrollable vividness, whereas mental imagery is typically voluntary, can be elicited by instruction in the absence of stimulation (except for after-images), and, to the degree indicated by Gordon's test of controllability of mental imagery, can have the content one wishes and be as vivid as one can achieve. Critically, synesthetes themselves distinguish between everyday mental images, which they can summon at will, and concurrents, which are firmly attached to specific stimuli (as illustrated for E.C. in Table 1). It is therefore of special interest that a mental image of a crayon color could, just like the crayon itself, elicit a taste in E.C. Yet this interaction is perhaps less clear-cut than it seems. Although E.C. did not know the names of the crayons at the start of the experiment, she had to learn them in order to generate appropriate imaginary colors when tested during the final session. To learn them, she had to see the crayon color while hearing its name ("periwinkle", etc.). Therefore, the route from mental image to concurrent may be via a learnt association with the name, rather than a direct effect. Evidence for a direct (non-verbal) link might be that E.C. experienced a brand-new sensation when presented with a compound of two or more distinct inducers at the same time. However, we have never found a compound that induces a new sensation; instead, we found that E.C. alternates between experiencing the concurrent due to each part of the compound, as illustrated in Figure 1. We therefore believe that the relation between mental imagery and synesthetic imagery needs further study.

An important further issue is whether synesthetic and real sensations have similar consequences. For example, Chiou et al. [19], when studying lexical-color synesthesia, found that real and synesthetic reds and greens, although equally vivid, differentially biased binocular rivalry. Real colors induce a localized sensory-level color-opponent (sensory) bias, whereas synesthetic colors induce a non-localized color-congruent bias, which Chiou et al. interpret as a cognitive effect. In the case of E.C., we can only report that real and synesthetic sweets, sours, and bitters seem similar to her, but unfortunately, we have no evidence to decide if this equivalence in experience translates into sensory substitution. For example, do synesthetic sweets adapt the way that real ones do? And if so, does the blocker affect their adaptations equally?

5. Conclusions

Our conclusions are based on data from a single, very rare, subject, and are speculative. This paucity of cases should not detract from the theoretical importance of E.C.'s synesthesia, however, in which—as is otherwise almost never the case—an all-or-nothing sensory manipulation could be applied to the concurrent. We hope that more cases of this sort can be found, so that the necessity or otherwise of an active bottom-up 'tonic' signal can be more firmly established.

Author Contributions: Conceptualization, C.C.-L.; Data curation, C.C.-L.; Formal analysis, A.R.; Writing—original draft, C.C.-L.; review & editing, A.R.

Funding: This research received no external funding.

Acknowledgments: We thank E.C. heartily for many hours of voluntary participation in the research.

Conflicts of Interest: The authors declare no conflict of interest.

References

1. Grossenbacher, P.G.; Lovelace, C.T. Mechanisms of synesthesia: Cognitive and physiological constraints. *Trends Cogn. Sci.* **2001**, *5*, 36–41. [CrossRef]
2. Bankieris, K.; Simner, J. Sound symbolism in synesthesia: Evidence from a lexical–gustatory synesthete. *Neurocase* **2014**, *20*, 640–651. [CrossRef] [PubMed]
3. Beeli, G.; Esslen, M.; Jäncke, L. When coloured sounds taste sweet. *Nature* **2005**, *434*, 38. [CrossRef] [PubMed]
4. Gendle, M.H. Word-gustatory synesthesia: A case study. *Perception* **2007**, *36*, 495–507. [CrossRef] [PubMed]
5. Jones, C.L.; Gray, M.A.; Minati, L.; Simner, J.; Crtichley, H.D.; Ward, J. The neural basis of illusory gustatory sensations: Two rare cases of lexical-gustatory synaesthesia. *J. Neuropsychol.* **2011**, *5*, 243–254. [CrossRef] [PubMed]
6. Ramachandran, V. The linguistic and cognitive factors associated with lexical-gustatory synesthesia: A case study. *Brain Cogn.* **2016**, *106*, 23–32. [CrossRef] [PubMed]
7. Simner, J.; Haywood, S.L. Tasty non-words and neighbours: The cognitive roots of lexical-gustatory synaesthesia. *Cognition* **2009**, *110*, 171–181. [CrossRef]
8. Simner, J.; Ward, J. The taste of words on the tip of the tongue. *Nature* **2006**, *444*, 23. [CrossRef]
9. Day, S. Some demographic and socio-cultural aspects of synesthesia. In *Synesthesia: Perspectives from Cognitive Neuroscience*; Robertson, L.C., Sagiv, N., Eds.; Oxford University Press: New York, NY, USA, 2005; pp. 11–33.
10. Yaro, C.; Ward, J. Searching for Shereshevskii: What is superior about the memory of synaesthetes? *Q. J. Exp. Psychol.* **2007**, *60*, 681–695. [CrossRef] [PubMed]
11. Coriat, I.H. A case of synesthesia. *J. Abnorm. Psychol.* **1913**, *8*, 38–43. [CrossRef]
12. Nikolinakos, D.; Georgiadou, A.; Protopapas, A.; Tzavaras, A.; Potagas, C. A case of color–taste synesthesia. *Neurocase* **2013**, *19*, 282–294. [CrossRef] [PubMed]
13. Craver-Lemley, C.; Mastrangelo, J. I can taste the figure, but not the ground: A case study of color-gustatory synesthesia. Presented at the 80th Annual Meeting of the Eastern Psychological Association, Pittsburgh, PA, USA, 3 April 2012.

14. Ward, J.; Simner, J. Lexical-gustatory synaesthesia: Linguistic and conceptual factors. *Cognition* **2003**, *89*, 237–261. [CrossRef]
15. Schroeder, J.A.; Flannery-Schroeder, E. Use of the herb Gymnema sylvestre to illustrate the principles of gustatory sensation: An undergraduate neuroscience laboratory exercise. *J. Undergrad. Neurosci. Educ.* **2005**, *3*, A59–A62. [PubMed]
16. Craver-Lemley, C.; Reeves, A. Is synesthesia a form of mental imagery. In *Multisensory Imagery*; Lacey, S., Lawson, R., Eds.; Springer Science Business Media: New York, NY, USA, 2013; pp. 185–206.
17. Intons-Peterson, M.J. Imagery paradigms: How vulnerable are they to experimenters expectations? *J. Exp. Psychol. Hum. Percept. Perform.* **1983**, *9*, 394–412. [CrossRef] [PubMed]
18. Yoshida, R.; Shigemura, N.; Sanematsu, K.; Yasumatsu, K.; Ishizuka, S.; Ninomiya, Y. Taste Responsiveness of Fungiform Taste Cells with Action Potential. *J. Neurophysiol.* **2006**, *96*, 3088–3095. [CrossRef] [PubMed]
19. Chiou, R.; Rich, A.; Rogers, S.; Pearson, J. Exploring the functional nature of synaesthetic colour: Dissociations from colour perception and imagery. *Cognition* **2018**, *177*, 107–121. [CrossRef] [PubMed]

Article

The Limiting Case of Amodal Completion: The Phenomenal Salience and the Role of Contrast Polarity

Baingio Pinna [1,*] **and Livio Conti** [2]

[1] Department of Biomedical Science, University of Sassari, 07100 Sassari, Italy
[2] Faculty of Engineering, Uninettuno University, 00186 Roma, Italy; livio.conti@uninettunouniversity.net
* Correspondence: baingio@uniss.it; Tel.: +39-39-2631-5770

Received: 28 April 2019; Accepted: 20 June 2019; Published: 24 June 2019

Abstract: In this work, we demonstrated unique and relevant visual properties imparted by contrast polarity in perceptual organization and in eliciting amodal completion, which is the vivid completion of a single continuous object of the visible parts of an occluded shape despite portions of its boundary contours not actually being seen. T-junction, good continuation, and closure are considered the main principles involved according to relevant explanations of amodal completion based on the simplicity–Prägnanz principle, Helmholtz's likelihood, and Bayesian inference. The main interest of these approaches is to explain how the occluded object is completed, what is the amodal shape, and how contours of partially visible fragments are relatable behind an occluder. Different from these perspectives, amodal completion was considered here as a visual phenomenon and not as a process, i.e., the final outcome of perceptual processes and grouping principles. Therefore, the main question we addressed through our stimuli was "What is the role of shape formation and perceptual organization in inducing amodal completion?" To answer this question, novel stimuli, similar to limiting cases and *instantiae crucis*, were studied through Gestalt experimental phenomenology. The results demonstrated the domination of the contrast polarity against good continuation, T-junctions, and regularity. Moreover, the limiting conditions explored revealed a new kind of junction next to the T- and Y-junctions, respectively responsible for amodal completion and tessellation. We called them I-junctions. The results were theoretically discussed in relation to the previous approaches and in the light of the phenomenal salience imparted by contrast polarity.

Keywords: amodal completion; shape perception; perceptual organization; depth perception; visual illusions

---

1. Introduction

The presence of a multiplicity of objects within the natural environment, the loss of one spatial dimension during the projection of the image on the retina, and the inverse-optics problem reveal a true challenge that all visual systems must face and solve. This is the occlusion among objects within a three-dimensional space. Occlusion is indeed a complex issue to the computations of visual surfaces since many components, surfaces, and parts of objects cannot have any counterparts in a retinal image. Furthermore, most of the visual objects are projected on the retina only as fragments, pieces, or parts of something, which have to be computed and of which the full shape must be "completed" by neural mechanisms.

The occlusion and the resulting completion can be seen in Figure 1, where a bunch of overlapping geometrical shapes is visible. Phenomenally, three circles, one square, one rectangle, one triangle, one pentagon, and one heptagon are partially perceived due to their reciprocal occlusion. Only the heptagon and the triangle are visible in their full shapes. This simple and spontaneous description demonstrates the prompt and effortless full completion of most of the fragments actually illustrated in Figure 1 and more clearly shown separated in Figure 2.

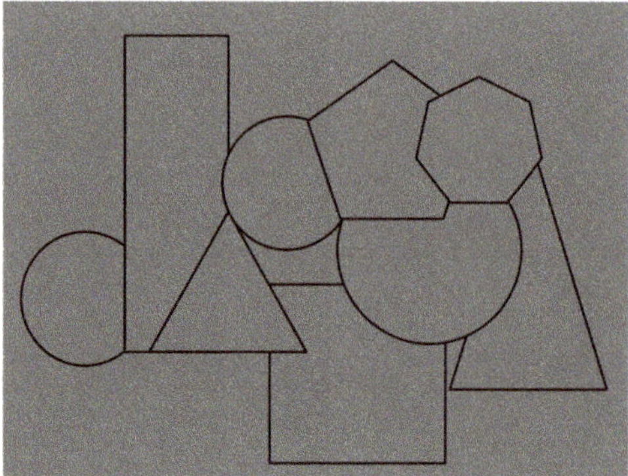

Figure 1. Amodal completion: the full and vivid completion of the visible portions of geometrical shapes behind other shapes.

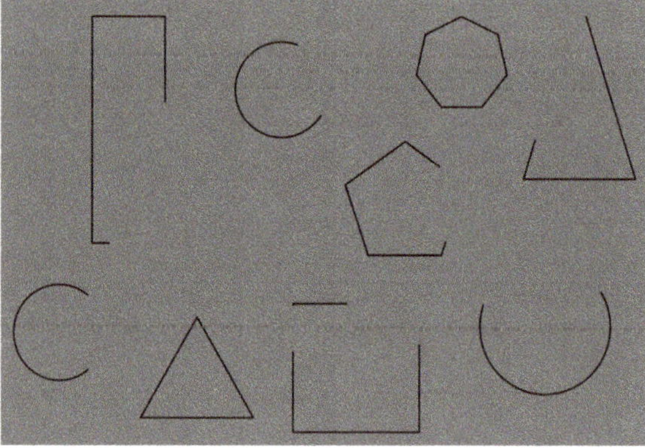

Figure 2. The visible portions of the amodal shapes perceived in Figure 1.

The completion of visible portions of each shape as a single continuous object is what is known as "amodal completion". Despite portions of boundary contours not actually being seen, the vivid outcome of a complete object unity is clearly perceived [1–8]. This is where the term "amodal" comes out. In summary, amodal completion is the intense sensory experience of completeness and unity of contours behind occluding objects (see also References [9–13]).

Amodal completion is likely the most common visual phenomenon and one of the most compelling problems of vision science. Psychophysical data demonstrated that several factors induce the perception of occlusion. They are mainly T-junction, asymmetry of Ts, good continuation, and closure [7,14–20]. Moreover, neurophysiological outcomes showed the role of neurons sensitive to amodal contours at higher visual levels [21–24]. Amodal completion is also related to processes of filling-in [7,25–27], thus showing that the visual world is not a mosaic of unconnected pieces of objects.

The outcomes, illustrated in Figure 1, are particularly suitable to be accounted for by Helmholtz's likelihood principle [28] and Gregory's "unconscious inference" [29,30]. They proposed that visual objects are similar to perceptual hypotheses postulated to explain the unlikely gaps within stimulus patterns according to what is perceived as the object that, under normal conditions, would be most likely to produce the sensory stimulation. In short, in Figure 1, partly occluded shapes are perceived because the other obvious possibilities, i.e., the fragments of Figure 2 abutting the occluding shapes, would require a coincidental and unlikely arrangement. Within the likelihood context, Rock [31,32] proposed the so-called avoidance-of-coincidences principle and claimed that the visual system tends to prevent interpretations elicited by coincidences. The visual system discharges coincidences tout court [33,34]. This is the case, for instance, of edges or junctions in one distal object that, through a specific view of a distal scene, accidentally coincide with edges or junctions in another distal object [33,35,36]. This principle can be proven selectively advantageous in the course of the phylogenetic development of the visual system.

This general idea of vision based on likelihood unconscious inference has been recently reconsidered in terms of probabilistic Bayesian inference [37–43]. It formally describes the optimal reasoning under uncertainty by specifying how to choose an outcome from a set of mutually exclusive hypotheses (Hs) on the basis of given stimulus patterns or data (D). According to Bayes' rule, the posterior p(H|D), indicating how likely H is for a given D (i.e., the relative degree of resulting belief for each hypothesis), is the result of the convolution between the prior p(H), namely how likely H is in itself, and the conditional p(D|H), i.e., how likely D is under H (the likelihood function: how well D fits H). Therefore, Bayes' theorem picks up the hypothesis that maximizes the posterior p(H|D).

In relation to Figure 1, the data are represented by the visual fragments projected on the retina and the hypotheses to be considered are the possible outcomes. Since the solution is undetermined, for example, due to the inverse-optics problem, Bayes' theorem computes a probabilistic decision aimed at choosing the outcome that becomes the conscious perception according to the incoming stimulus pattern. This is supposed to occur by maximizing the posterior distribution. Therefore, the likelihood function, i.e., p(D|H), models aspects of optics and projection on the retina, while the prior, i.e., p(H), models the constraint and prior assumptions on the structure of the environment necessary to solve underdetermination.

In short, by reconsidering Figure 1 in terms of Bayes' rule as stated in Helmholtzian likelihood principle and in the avoidance-of-coincidences principle, the most likely interpretation of Figure 1 is expected to be the set of full shapes previously described. This is also assumed to be the simplest solution.

Phenomenal simplicity refers to the notion that the visual system tends to extract a maximum of regularity. This idea is based on the simplicity–Prägnanz principle of Gestalt psychologists, who consider the visual system, like every physical system [44], as aimed at finding the simplest and the most stable organization consistent with the sensory inputs [45].

Simplicity and likelihood have been considered competing theories [31,46–50] that explain perceptual organization. The main difference is related to the fact that simplicity is based on a general principle of economy while likelihood is based on probability. In spite of this basic difference, they can be seen as two ways of considering the same visual process (see Reference [51]). As a matter of fact, the visual object that minimizes the description length is the same one that maximizes the likelihood. In other terms, the most likely hypothesis about the perceptual organization is also the outcome with the shortest description of the stimulus pattern.

Phenomenally, the most basic cues for amodal completion responsible for simplicity and likelihood are T-junctions [7,52–54]. The role of T-junctions within amodal completion can be explained by Occam interplay between priors and conditionals. Feldman [39] also suggested that T-junctions are mainly cues for segmentation since they, first of all, have to be segmented into visible parts of two different surfaces.

It should be pointed out that the most elementary condition of amodal completion is the figure–ground segregation, according to which a visual object partially occludes its back side and, at the same time, a portion of the background. Rubin [55,56] first studied figure–ground segregation as an essential process to the existence of visual objects by studying general principles of figure–ground segregation assumed as the atoms of a more general grammar of phenomenology of vision. Rubin's main principles are surroundedness, size, orientation, contrast, closure, symmetry, proximity, convexity, and parallelism.

In Figure 1, the junctions among fragments are clearly related to their completion in object unities; however, not all of them are T-junctions. This is not an issue since all kinds of junctions can be reviewed in terms of the Gestalt principle of good continuation. For instance, T-junctions tend to be seen as visible parts of two different surfaces because the orientation of the vertical component of the "T" is dissimilar (orthogonal) from one of the two halves of the horizontal component, which is the best possible (good) continuation of the other half of the horizontal component. Both halves have the same orientation, the best or good possible continuation. This implies that T-junctions can be considered special conditions among many others, with intersecting contours not necessarily orthogonal of a Gestalt good continuation, that is *ipso facto* a generalization and a phenomenal explanation of the role of the junctions in eliciting amodal completion in Figure 1 [1,2,8,14–18]. Another kind of special junction is Y-junctions, eliciting juxtaposition and tessellation of surfaces (see next sections). In short, the bifurcation designed by the Y-junctions, in terms of good continuation, elicits two possible directions in equilibrium or equal probability. Therefore, the good continuation is not stopped by any boundaries as in T-junctions and is mainly responsible for the amodal continuation and completion. In this work, we proposed a new kind of junction, through limiting cases that we called I-junctions (see next sections). It is worth noting that T- and Y-junctions and, more generally, the good continuation can be read as a special case of similarity/dissimilarity among orientations of intersecting contours. This remark will be reconsidered below by matching this similarity due to orientations with the one elicited by contrast polarity.

There is a further phenomenal attribute that is worth highlighting and that will be explored in the next sections. It is the amodal continuation. In short, it is the apparent and vivid outcome of continuation behind an occluding figure. As such, it is evidently related to amodal completion; nevertheless, it can be considered as operating more locally and partially independent from the more global-shaped organization due to amodal completion. For the time being, we direct reader's attention to Figure 1, where the difference between amodal completion and continuation can be noticed. In Figure 1, the square underneath and the large triangle on the right appear distorted, respectively, in the square's top-left corner and in the upper-corner of the triangle. These distortions or deformations are related to the continuation of the two fragmented sided of the figures that, continuing in their orientation, do not meet at the right point of shape regularity (for a more focused work on this matter, see Reference [7]). To better appreciate these effects, compare Figure 1 with Figure 3, where the complete, geometrical, full, and transparent overlapping of the inner shapes of Figure 1 is illustrated.

Figure 3. The invisible geometrical full and transparent overlapping shapes of Figure 1.

Given the segmentation induced by T-junctions, to perceive the intense sensory experience of completeness and unity of the amodal completion, it is necessary to perceive an occluding object and, thus, the perception of illusory depth. The continuous and smooth edges usually belong to the occluding object, whereas the intersecting (differently oriented) edges belong to the occluded object according to the unilateral belongingness of the boundaries [55,56] and the border ownership principle [7,28,57–64]. This is also the case of Figure 1, where the multiplicity of complete objects are perceived as placed at different depths and with some of them partially overlapping others.

This is also the case of the well-known Kanizsa's triangle, where brightness enhancement and illusory contours are seen in the absence of a luminance or color change across the contour. Three black sectors and three angles, arranged respectively along the vertexes and sides of a virtual triangle, are perceived as three black disks and an outlined triangle in depth behind a triangle with clear boundary contours and brighter than the white background.

Based on Gestalt theory, Kanizsa [9,10] suggested that the necessary factor for the formation of the illusory figure is the presence of incompletenesses, or open figures, inducing amodal completion and closure processes that "create" complete perceptual elements behind a partially occluding illusory triangle.

Since amodal completion induces the formation of complete shapes and depth perception, the question that spontaneously attracted most scientists was "What is the role of amodal completion in shape formation?" [3–5,7,9,15,65–67].

The main target of this question is the shape completion of the partially occluded object, namely the shape that amodally completes the visible fragments. This question assumes the amodal completion as the cause of the formation of shape. Roughly speaking, the process of amodal completion is considered the main one responsible for the illusory depth and shape completion of the occluded object. Since the amodal shape is the object hypothesis of the perceived occluded object, what remains to be explained is the precise shape of the amodal objects. Simplicity and likelihood approaches are aimed at explaining how the occluded object is completed, its shape and information load, and how contours of partially visible fragments are relatable behind an occluder.

Complementary to this approach, there is another one that considers amodal completion not as the cause or the starting point but as the resulting effect, i.e., the end-point of the visual segmentation chain. The questions are, then, "What is the role of shape formation and perceptual organization in inducing amodal completion? Again, what are the perceptual conditions that elicit the segregation of occluded and occluding objects and, finally, amodal completion? Moreover, what is the role of the local contours, junctions, and termination attributes in eliciting the phenomenon of amodal completion?"

The answers to these questions allow us to understand the perception of illusory depth and the emergence of occluding and occluded objects. Within this perspective, amodal completion is reconsidered in terms of shape formation and, as such, reduced to elementary and more general principles of grouping and figure–ground segregation. Within the last question, amodal completion is considered as a visual phenomenon, while in the previous question ("What is the role of amodal completion in shape formation?"), it is assumed as a process aimed at explaining the perceived amodal shape.

On these bases, the main purpose of this work is to answer the last questions and to investigate more in detail amodal completion as a phenomenon not taken for granted but as the final outcome of perceptual organization: the result of upstream dynamics of shape formation and grouping principles. We assumed that this perspective could be a good candidate to test Helmholtz's likelihood principle, the avoidance-of-coincidences principle, and Bayes' framework.

The contrast polarity, with its related similarity/dissimilarity outcomes, is the main grouping and ungrouping attributes used in the next sections to explore amodal completion as a visual phenomenon.

Recently, contrast polarity has been demonstrated as an effective visual attribute in imparting strong grouping effects within a pattern of stimuli [68–71]. It can put together otherwise segregated elements or disrupt joined components. Figures 4 and 5 show respectively the grouping and ungrouping properties induced by contrast polarity in favor or against the good continuation and T-junctions of Figure 1.

Figure 4. When the contrast polarity plays synergistically with other factors (T-junctions and good continuation), amodal completion and depth segregation are more salient than those perceived in Figure 1.

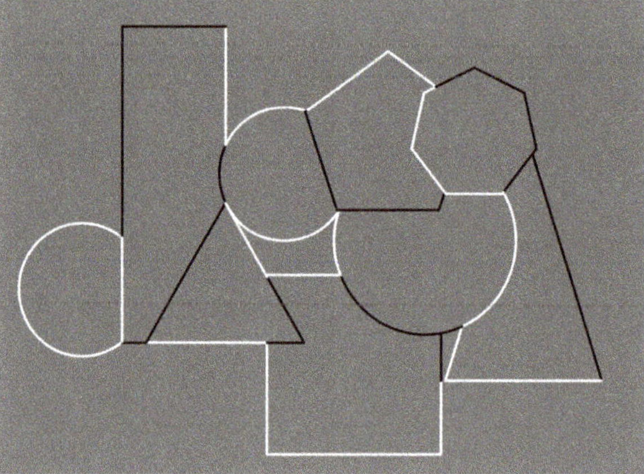

Figure 5. The results of Figure 4 are now partially and slightly disrupted, parceled, and camouflaged due to the contrast polarity pitted against T-junctions and good continuation.

Without going into detail, in Figure 4, the salience of both depth segregation and amodal completion of partially occluded shapes are stronger than the ones illustrated in Figure 1. Under these conditions, contrast polarity plays synergistically with good continuation and T-junctions. On the contrary, in Figure 5, since the contrast polarity is now pitted against the same factors, both amodal completion and the unity of the outcomes of Figure 4 are partially disrupted, parceled, and camouflaged.

The new conditions studied in the next sections will deepen the role and strength of contrast polarity when it is pitted against or in favor of other grouping principles of perceptual organization.

2. Materials and Methods

2.1. Subjects

Different groups of 20 undergraduate students participated in each experiment described in the next sections. About 20% of the subjects had some basic knowledge of visual illusions, Gestalt psychology, and amodal completion, and the others were totally naive both to the stimuli presented here and to the purpose of the experiments. They were about 50% male and female, and all had normal or corrected-to-normal vision.

2.2. Stimuli

The stimuli were the figures, illustrated in the next sections, mainly based on a cross geometrically composed of five adjacent squares or a cross made up of two-centered and intersected orthogonal rectangles of equal size. Further stimuli were juxtaposed, overlapped, and isolated polygons. The stimuli were displayed on a 33-cm color CRT monitor (Sony GDM-F520 1600 × 1200 pixels, refresh rate 100 Hz, Sony, Tokyo, Japan) driven by a MacBook computer with an NVIDIA GeForce 8600 M GT in ambient illumination provided by a Osram Daylight fluorescent light (250 lx, 5600 K). The overall sizes of the figures were approximately 5 degrees. The luminance of the white elements was 122.3 cd m^{-2}. Black components had a luminance value of 2.6 cd m^{-2}. The gray background luminance was 62.5 cd m^{-2}, about halfway between the white and black line components.

Viewing was binocular in the frontoparallel plane at a distance of 60 cm from the monitor.

2.3. Procedure

Given the complexity of the questions we are going to answer, we addressed related issues by showing novel and possibly fruitful conditions based, first of all, on self-evident perceptions and, secondly, on experimental phenomenology [58,59,71–75], including qualitative observations under controlled conditions rather than psychophysical results. Our purpose was to answer the questions by means of phenomenal outcomes that, being as strong as possible, could isolate the qualitative role of perceptual principles useful in testing the effectiveness of predictions and theories. Therefore, the stimuli have been designed to serve as limiting conditions or *instantiae crucis* (crucial instances) aimed at testing general theoretical statements through detailed phenomenal properties. Each of our experiments can be, therefore, read as an *experimentum crucis* (see also References [45,76–78]).

The procedure was twofold in line with the classical ones used by Gestalt psychologists (see also References [9,10,45,79–84]). The first is the phenomenological free-report method, through which naive subjects were asked to report anything they see in the following series of visual stimuli. The second is a more quantitative method, according to which subjects were instructed to rate (in percent) the descriptions obtained in the phenomenological experiments.

Phenomenological task—The task of the subjects was to report spontaneously what they perceived for each stimulus by giving, as much as possible, an exhaustive description of the main visual outcomes. The descriptions were judged by three students of linguistics that were totally naive to the hypotheses in order to get a fair representation of the responses of the subjects. All reports occurred fast and spontaneous. Observation time was unlimited, with the observers looking at the stimuli during their report.

Participants could make free comparisons, add comments as afterthought, and view the displays in different ways and from different distances. Subjects could receive suggestions/questions of any kind, such as "What is the shape of each component? What is the whole shape?" All the variations and possible comparisons occurring during the free exploration were noted by the experimenter. They could also match the stimulus with every other one they (or the experimenter) considered appropriate. This degree of freedom is in line with experimental phenomenology and aimed at more stable outcomes. The selection of stimuli could involve opposite conditions, controls, and possible comparisons between stimuli. The stimuli were presented randomly to minimize biases and past experience.

Scaling task—Subjects were asked to rate (as percent) the main descriptions resulting from the previous phenomenological task. At this stage, new groups of 20 subjects were asked to scale the relative strength or salience (in percent) of the main outcomes. Their task was literally: "please rate whether this statement (e.g., "five-squares juxtaposed and placed on the same depth plane" or "two-rectangles placed at a different depth") is an accurate reflection of your perception of the stimulus on a scale from 100 (perfect agreement) to 0 (complete disagreement)". We report below descriptions of which the mean ratings were greater than 85 across all experiments (about these procedures, see also References [58,59,68,82,85]). Critically, the statements rated were based on a careful analysis of previously obtained spontaneous descriptions, so the subjects were not being forced to rate appearances that no one had reported before.

In the following sections, the reported descriptions are incorporated within the text to aid the reader in the stream of argumentations.

3. Results

3.1. Contrast Polarity, Similarity, and Good Continuation

The starting question is "What is the role of shape formation and perceptual organization in inducing amodal completion?" To answer this question, the main purpose is studying the complexity of the perceptual organization through contrast polarity, which is useful to understand the complexity of amodal completion and to test the effectiveness of the theoretical approaches previously described. The role of contrast polarity will be investigated in relation to T-junctions, good continuation, and regularity.

The basic condition for the first set of stimuli is the Greek cross-like shape with all arms of equal length and equal sides illustrated in Figure 6. The cross is composed of five adjacent squares. This geometrical description becomes challenging for our purposes when it is compared with the following phenomenal report: a cross composed and made up of two centered and intersected orthogonal rectangles of equal size. The rectangles are overlapping, defined only in their boundaries or perimeter (outlined only) with the inner edges totally transparent or empty.

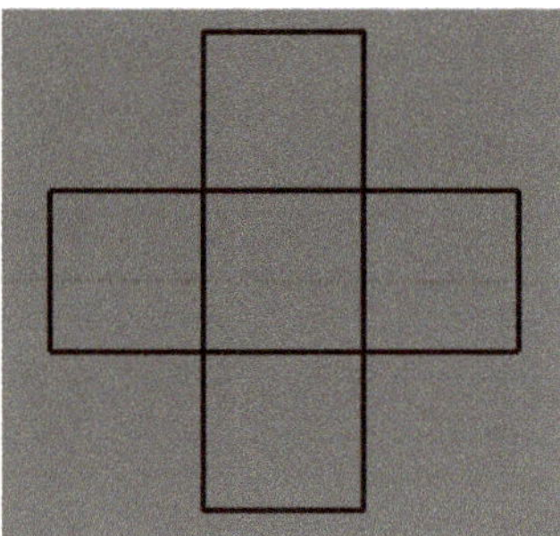

Figure 6. A cross geometrically composed of five adjacent squares or a cross made up of two-centered and intersected orthogonal rectangles of equal size?

In terms of simplicity and Bayes' inference, this description is privileged in relation to the cross composed of five juxtaposed squares. From the perspective of Bayes' inference, the geometrical five-squares solution, although unlikely, is processed anyway.

The phenomenal outcome is privileged, primarily, because the number of elements is minimized: two rectangles vs. five squares. Secondly, the ambiguity related to the number of elements is also reduced. In this regard, we asked our subjects to answer the following questions, "Now, try to perceive adjacent squares; what is the number of squares?" Since we have already answered this question, namely "five", the question is apparently trivial. Nevertheless, looking more carefully at the stimulus, the number of squares could be four or five. In fact, eleven out of twenty subjects answered "four". This implies that they perceived the central square not as a filled surface but as an empty space. This result is mainly due to Rubin's unilateral belongingness of the boundaries (border ownership). Indeed, to create the pattern of Figure 6, the juxtaposition of the four squares at the angles is sufficient. The central square can be implicit, that is, given by construction. Moreover, phenomenally, the boundaries (i.e., the four surrounding squares) are perceptually the best candidates to describe the whole shape (the cross). The nine remaining observers perceived five filled squares.

Actually, the geometrical and phenomenal perspectives can be considered as two different ways of watching the stimulus, that is locally or globally, and under these conditions, there is no ambiguity: respectively, five squares and two rectangles. Furthermore, the geometrical way reveals a phenomenal tessellation, while the phenomenal way highlights an overlapping and intersection outcome.

There is another more effective phenomenal quality favoring the two-rectangles solution. This is the good continuation principle that makes each longer contour appear as unique and seamless and as a unity. It prevents the longer side of the two orthogonal rectangles from being broken in three pieces and then brings about the adjacent-squares solution. The breakdown of the longer contours is the opposite of the tendency of the elements with more similar orientations to be grouped together. Actually, each intersection is a bifurcation and the visual system is expected to compute a Bayesian decision. In phenomenal terms, the result of Figure 6 states, "All else being equal, whenever there is a bifurcation and changes of direction, they can be prevented (minimized), unless otherwise stated, they do not change". An algorithmic way to code the principle of good continuation could be "Avoid (or minimize) changes" or "Follow the minimum changes". By writing "unless otherwise stated", we are referring to the new conditions that are going to be introduced (see next figure).

There is a further complexity to be considered in the two-rectangles solution: (i) the perceived overlapping of the rectangular wires and (ii) which is in front and which is in the back. The perceived order of superimposition in Figure 6 is perceived as reversible, i.e., alternately, the horizontal and vertical rectangles can be perceived one in front of the other. This problem has been studied first by Petter [86], who discovered an interesting effect concerning conditions where a full black irregular pattern may be seen as made up of independent surfaces separated in depth and delineated by illusory contours in the area of apparent intersection and stratification. Petter also studied patterns formed by transparent surfaces or formed by outlined surfaces similar to the ones illustrated in Figure 6. He suggested that the perceived stratification occurs according to a general rule stating that the surface with the shorter contours, placed in the region where the surfaces look superimposed, has a greater probability of appearing in front of the other surface [9,87,88]. Since the length of the contours is equal, then the expected order of superimposition should be and it is reversible. Shipley and Kellman [87,88] also demonstrated that completion occurs not only for two-dimensional shapes but also for one-dimensional line figures. Moreover, inferred shapes can happen for curvilinear as well as linear elements.

Summing up, while the geometrical five-squares are seen as juxtaposed and all placed on the same depth plane, the two-rectangles are perceived as placed at a different depth. It seems that the minimization of the number of elements implies the emergence of depth. The reduction of the number of components implies the addition of a new dimension, useful to explain the reduction.

These preliminary descriptions of Figure 6 seem comprehensive and complete. However, they are just two among a plethora of possible results not easy to calculate, although, within Bayes' inference, they are all taken into account and computed. Even with a simple stimulus like this, the computation could entail an exponential algorithmic complexity.

Moreover, since the likelihood of all the combinatorial possibilities, except the two here considered, is about zero, then a fortiori it is an algorithmic waste to compute all the possible combinations. Given the fast and effortless answer of the subjects, it seems that the visual system knows exactly what to see without any need to compute combinations. This is due to the good continuation that preserves the visual system from computing futile and unlikely combinations. It is worthwhile to report that when subjects were asked to make an effort to perceive other possible combinations, they obstructed the task, stating that they are unable to calculate and perceive them. In Figure 7, some of these combinations are highlighted through the contrast polarity.

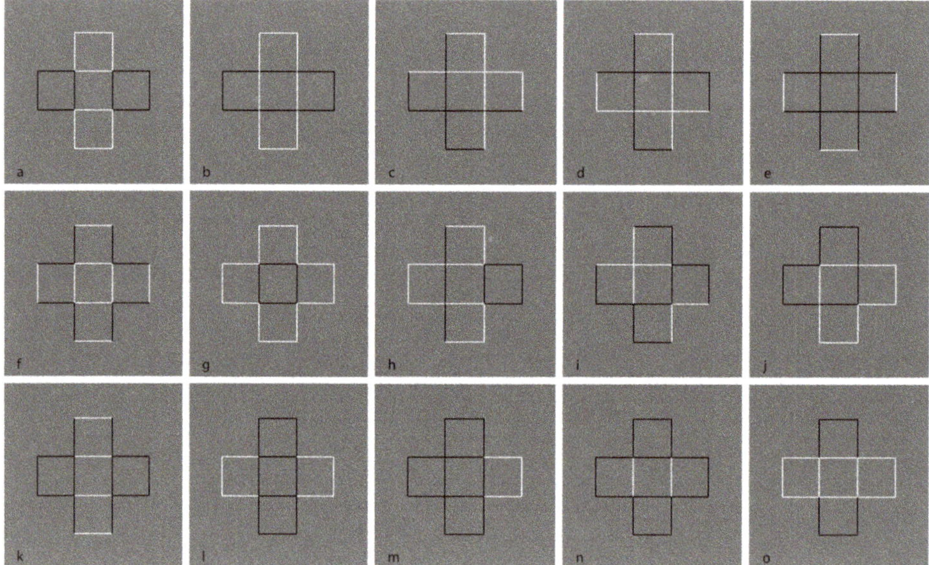

Figure 7. Conditions demonstrating the dominance of the contrast polarity over good continuation, T-junctions, and regularity.

Another point that is worth being considered is related to the simplicity/Prägnanz of the previous results. More particularly, among the plethora of possible combinations, we can ask if there are other combinations that could be simpler than the two-rectangles outcome. Theoretically, the number two can be reduced to one if only the cross could be perceived. Actually, this is what has been reported, but phenomenally this answer is slightly but significantly different from a full perception of a cross: "a cross made up of two rectangles". As a matter of fact, this is not a cross, i.e., a secondary and indirect or virtual object made up of two primary rectangles. The point remains open, but the answer might be possible among the phenomenal results of Figure 7.

If in Figure 6, T-junctions and good continuation are the main principles involved (closure should also be considered, but within this stimulus, they are balanced), in Figure 7 the similarity principle elicited by contrast polarity is now pitted in favor or against the previous results and the principles involved. More in detail, in Figure 7a, both T-junctions and good continuation are broken apart and the unique phenomenal outcome is four squares, two white and two black. The inner squared region is empty, as previously described, but with the difference that, in Figure 7a, this solution is much stronger than the one in Figure 6 and totally dominant against all the others, that are unlikely.

In Figure 7b, the overlapping two-rectangles solution is restored and made more salient by the contrast polarity. Under these conditions, the white rectangle has been reported as more vividly perceived popping out in front of the black one. In Figure 7c,d, the two rectangles are perceived with a clear bulging and volumetric effect. While in Figure 7c, both rectangles appear as convex, in Figure 7d the horizontal rectangle is seen as concave.

These volumetric effects can be explained on the basis of the assumption that the direction of the light source is from above. Actually, the qualitative volume interpretation of a visual object comes from the light-source constraint. If the light source lies in the concave side of a cast shadow contour of the rectangle, it should be perceived as concave; otherwise, the rectangle should be convex [73,89–98]. The convexity/concavity of the rectangle is determined by whether the shadow contour bends away from the casting edge or toward it. The assumption or Bayes' prior [99,100] that a single light source comes on the right side from above disambiguates the qualitative shape of the rectangles to be convex or concave.

Despite this explanation, what is puzzling within these two figures is the fact that the rectangles show a clear volumetric effect and, at the same time, they are transparent, empty, or outlined only. Since the volumetric effect is assumed on the basis of the light source, the presence of outline only rectangles contradicts the Bayesian prior. Alternately, it can be suggested that the similarity and grouping induced by the contrast polarity precede the assumption of the light source; thus, the contradiction is only apparent since it occurs at a higher level after the shape formation and organization.

In Figure 7e, the white sides of the rectangles, placed on the extreme arms of the cross, highlight the visibility of the cross, which emerges more and more clearly in Figure 7f,g. The similarity/dissimilarity among components weakens the good continuation and highlights the external boundaries of the cross. Again, the role of the contrast polarity dominates the other principles. The cross is now seen distinctly and directly, while in Figure 7, it appears indirectly as the result of a secondary and holistic visual process related to the arrangement of the external boundaries. Moreover, since Figure 7g shows two objects (the cross and the square) like Figure 7b (two rectangles), the question begs, Which is the outcome to be privileged in terms of simplicity/Prägnanz? The answer to this question is per se very difficult if the role of contrast polarity is not appropriately considered. However, by invoking the contrast polarity, the simplicity/Prägnanz fails when the following conditions are examined.

The grouping strength of contrast polarity is better tested in Figure 7h–j, where new figures, invisible in Figure 6, pop up. The contrast polarity now also plays against the regularity and, more importantly, against simplicity/Prägnanz. In Figure 7h, a straight vertical line, a zig-zagged Greek-like pattern, and a square are perceived. Figure 7i shows an irregular grouping difficult to describe both phenomenally and geometrically, as reported by most observers. In Figure 7j, two equal arrowhead-like shapes pointing in opposite directions have been reported.

Given the similarity/dissimilarity and grouping/ungrouping effects of the contrast polarity within the same pattern of stimuli, the amodal completion can be elicited even when the good continuation is expected to be more in favor of the contrast polarity and, thus, without the need for amodal completion (see Figure 7k–l). In Figure 7k, a horizontal rectangle is amodally perceived behind a vertical rectangle made up and segmented by horizontal white contours. This result entails that the white contours belong to the vertical rectangle, although on the basis of the good continuation, they are expected to group with the horizontal amodal rectangle. Figure 7l shows a strong amodal completion of the partially visible components of the horizontal rectangle, placed behind the vertical segmented one. In Figure 7m, only the right white portion of the figure tends to complete amodally behind the black remaining part of the figure.

In Figure 7n, the two upper and lower portions tend to complete amodally behind the horizontal segmented rectangle. Finally, Figure 7o demonstrates the condition, complementary to Figure 7l, where a horizontally segmented rectangle partially occludes a vertical black rectangle amodally completing behind it.

It is worth noting that, in these new conditions of amodal completion, the rectangles are not perceived only in their perimeter or boundaries. They are not transparent but appear filled and opaque; otherwise, there would not have been any amodal completion. Actually, the continuation of the black vertical contours of Figure 7o appear to complete amodally behind the white contours. This implies a new kind of amodal completion without T-junctions but with I-junctions, where the continuation of contours is behind a contour with the same orientation. The two contours, although with the same orientation, are perceived as two different contours. Further and stronger conditions of this new case of amodal completion will be presented in the next sections.

From these results, the contrast polarity manifests both grouping and ungrouping effects likely due to their salience and visibility, due in turn to the largest amplitude of luminance dissimilarity imparted. In other terms, contrast polarity operates by highlighting components that, on the basis of the resulting similarity and dissimilarity, groups and ungroups accordingly.

The phenomenal results of these stimuli clearly demonstrated the domination of the contrast polarity against good continuation, T-junctions, and regularity. Since they are principles closely related to amodal completion, their role in relation to the contrast polarity is worthy of being more deeply examined to better understand amodal completion in itself and to answer our starting question: What are the role of shape formation and perceptual organization in inducing amodal completion?

As a final remark, Figure 7 questions the effectiveness of simplicity/Prägnanz, information load, and Bayes' inference. Unless we do not consider the contrast polarity as a constraint or as a prior, Bayes' inference cannot easily explain these conditions. Moreover, our results cast doubts also on Kolmogorov [101] complexity, which is a natural extension of Occam's razor. It subsumes the principle of the minimum description length, according to which, given stimuli and a choice of different possible outcomes, it pushes to choose the outcome such that the description of the outcome plus the conditional description of the data is as short as possible. Some of our results, as previously demonstrated, do not conform to this general assumption. Moreover, the high number of possible outcomes previously described represents already a clear weakening of the Kolmogorov complexity. Though many or all the figures can contain a shorter description of the data, it seems that only the contrast polarity chooses the final result, without taking into account either Kolmogorov complexity or the other related principles. We suggest that it is exactly the popping out imparted by the reversed contrast that is the main factor responsible for the described phenomena. The salience and visibility, derived by the largest amplitude of luminance dissimilarity imparted by contrast polarity, precedes any holist or likelihood organization due to simplicity/Prägnanz and Bayes' inference.

The next step is to test the contrast polarity in more classical conditions used to study amodal completion.

3.2. Contrast Polarity and Amodal Completion

The question of this section is as follows: What is the role of contrast polarity in eliciting amodal completion? The basic stimulus of the following new set of conditions is now shown in Figure 8a, where a classical example of amodal completion is illustrated and spontaneously perceived as a vertical rectangle occluding a horizontal rectangle partially visible on its left and right sides. In Figure 8b, the previous visual organization is strengthened and described as more vivid than the one in Figure 8a.

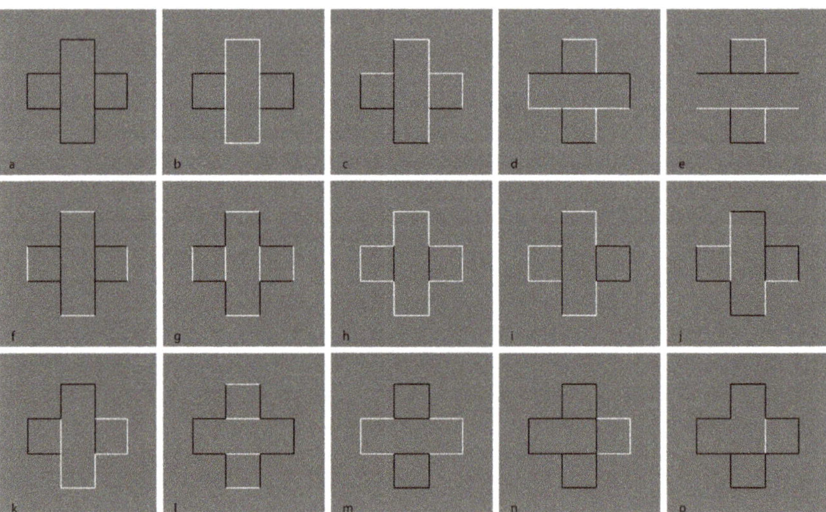

Figure 8. The dominance of the contrast polarity over good continuation, T-junctions, and regularity is also demonstrated within a classical example of amodal completion.

Figure 8c,d show the volumetric effects, previously described, without the antinomic phenomena induced by the overlapping or intersection of outlined only rectangles. The prior light source from above seems to be effective in both conditions, revealing clear convex (Figure 8c) and convex vs. concave (Figure 8d) results. However, a novel issue resides in Figure 8d, where, despite the T-junctions and the contrast polarity pitted in favor, the amodal completion is absent. Actually, the two upper and lower squares are perceived as bulging closer to the observer than the horizontal concave rectangle. Theoretically, there is not a reason or a constraint in terms of T-junction or border ownership that should prevent or inhibit the amodal completion of bulging objects behind a concave surface. However, there is a physical constraint that should prevent the amodal completion of bulging objects behind a concave surface. It is because, in terms of depth, the concave surface (horizontal rectangle) is further away from the observer than the background and the convex surface (top and bottom squares) is closer to the observer. Therefore, as long as the edges of a convex vertical rectangle and a concave horizontal rectangle are at the same depth as the background, a convex vertical rectangle cannot exist behind a concave horizontal rectangle.

In Figure 8c, two bulging objects are perceived one behind the other. Why should a bulging object not take place behind a concave one? The main point is likely the fact that the bulging squares are perceived closer to the observer than the concave rectangle. This is clearly visible. However, the border ownership and the good continuation show, still clearly, that the side of the square, geometrically in common with the rectangle, is perceived as belonging to the rectangle. As a consequence, the longer sides of the rectangle should be and, indeed, are perceived closer to the observer than the squares.

Furthermore, the squares should be perceived as behind and, since they are incomplete with a missing side, they are therefore expected to complete amodally behind the rectangle. This is not the case. The squares appear in front even if they appear behind the longer sides of the rectangle, which are perceived as behind although the sides are seen in front of the squares. This is a true puzzle or a visual paradox, where different scales and locations of the visual field reveal different and antinomic results that cannot be solved through a unique logic or object hypothesis.

On these bases, both Helmholtz's likelihood principle and Bayes' inference are clearly questioned since they use the same global logic in every location and on all size scales of the stimulus. Actually, by removing the two small sides of the rectangle, the antinomy is less strong and the two squares can more easily complete amodally behind the two horizontal contours (Figure 8e). This result is stronger if the two long horizontal sides are not perceived as belonging to the same objects but just as two independent segments. The spatial separation between the two upper and lower parts increases the strength of this outcome.

In Figure 8f, the closure of the components of Figure 8a with white contours improves their figurality and, as a consequence, the amodal completion of the visible left and right elements in a rectangle. In Figure 8g, turning the intersecting contour of Figure 8f from black to white breaks the amodal continuation and induces the pop out of the two lateral squares as complete objects placed in front of the vertical rectangle. Under these conditions, a further case of I-junction, even more salient than those previously described, is perceived. Figure 8h demonstrated that the amodal continuation along the I-junction depends not only on the contrast polarity but also on the arrangement of white and black contours and by the closure principle. The white and black contours segregate in two components, eliciting a white cross with two black contours closing the two horizontal arms but not creating any depth segregation among the regions.

A clear depth segregation is instead visible in Figure 8i, where a vertical rectangle shows, on its left and right sides respectively, an amodal white square-like shape continuing behind its longer side and a complete black square in front of the rectangle. The two lateral shapes reveal a T-junction on the left and an I-junction on the right.

More complex figures to be described are demonstrated in Figure 8j,k, revealing, firstly, an irregular and confusing arrangement of components without any global logic or intelligible likelihood and, secondly, local groupings with local amodal completion or superimpositions separated, independent, and unconnected. Locally, they group and ungroup on the basis of the dominant contrast polarity rule of similarity/dissimilarity.

Figure 8l was described similarly to Figure 8g but with the two squares placed on the upper and lower boundaries of the rectangle and showing again the I-junctions. The effect of the I-junctions is even stronger in Figure 8m, where the white rectangle is clearly perceived as placed and continuing amodally behind and exactly along the sides of the two squares on the top and on the bottom. What is completed is not the surface or area or shape but just the continuous contours of white and black lines.

The last images of Figure 8n,o show conditions that are, again, difficult to describe and locally obey the contrast polarity grouping and ungrouping as previously stated for Figure 8j,k.

3.3. Contrast Polarity and Petter's Effect

A special case of amodal completion is Petter's effect [6,86], previously introduced. This effect is seen when a homogeneous black irregular pattern is perceived as made up of independent surfaces separated in depth and delineated by illusory contours in the area of apparent intersection and stratification (Figure 9). Petter [86] discovered that, the larger surface, the region with simpler boundaries and in motion appear, respectively, in front of the smaller, of the one with more complex boundaries, and of the static one.

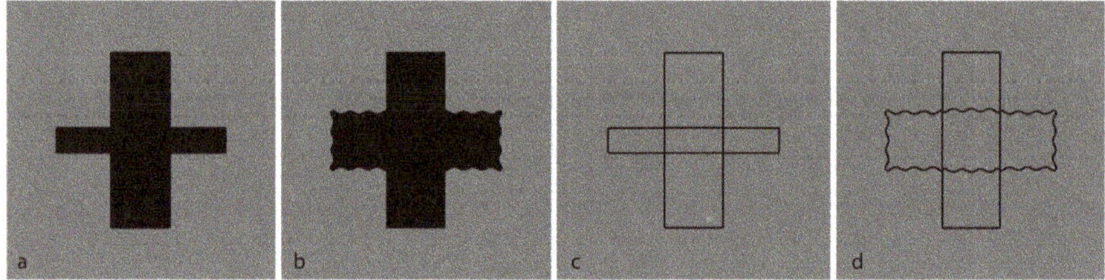

Figure 9. Petter's effect: the larger surface (**a**) and the region with straight boundaries (**b**) appear in front of the smaller one and of the region with undulated boundaries.

In Figure 9a,b, two examples of Petter's effect are shown, where the larger rectangle is perceived in front of the thinner one (Figure 9a) and the vertical rectangle with straight boundaries is seen in front of the one with undulated contours (Figure 9b). Figure 9c,d demonstrates that Petter's effect can also be perceived with outlined patterns although with less strength.

Petter also suggested a general rule that states that the region with the shorter contours intersecting the one where the surfaces look superimposed has a greater probability of appearing in front of the other surface because it requires a lower energy expenditure. This idea is apparently in accordance with Gestalt theory and with the general "minimum principle" of Prägnanz [45] (Koffka, 1935), suggesting the tendency of neural activity toward minimum work and minimum energy, which spontaneously elicits self-organization toward simplicity. This idea was related to the dynamics of physical systems that naturally converges on a state of minimum energy [19,20,44,79,80,102,103].

Stoner and Albright [104] proposed that Petter's rule is a consequence of the heuristic and stated that larger figures should be seen in front as a result of perspective projection. Therefore, a close object has a larger retinal projection. Nevertheless, other studies demonstrated that the length of interpolated contours can vary independently from the figure size [105]. These results weaken the connection of minimum principle with likelihood and, ultimately, with simplicity/Prägnanz. The supposed connection can be weakened even more by merely observing Figure 9a–d. The question is why should the solution that is larger in front be simpler than that which is smaller or thinner behind? Moreover, is it more likely that objects with straight boundaries are perceived in front of objects with undulated contours?

These questions become more intriguing by introducing the contrast polarity within the play of Petter's rule. The main argument is if Petter's rule is working within a pattern of elements, then it should reduce or amplify the effect of the grouping induced by contrast polarity once the two rules play against the other or synergistically.

To explore this argument, the same set of stimuli illustrated in Figure 7 has been redrawn in Figure 10 by introducing Petter's effect. The results can be summarized as follows. By comparing each stimulus of Figure 10 with its counterpart of Figure 7, from Figure 10b to Figure 10e, where contrast polarity is in favor of Petter's effect, the larger rectangle is clearly perceived in front of the thinner one. However, in all the other stimuli, where the contrast polarity breaks the good continuation and the two-rectangles organization, the resulting effects previously described for each stimulus of Figure 7 are in Figure 10 more vivid and stronger. This is likely related to the proximity principle among nearest components, segregated and grouped through the contrast polarity, that favors their emergence as figures segregated from the furthest. For example, in Figure 10a, the two small lateral rectangles are perceived more clearly segregated from the large ones, since they are smaller with nearer sides. Being smaller with close elements elicits a stronger figure–ground segregation as predicted by Rubin's principles.

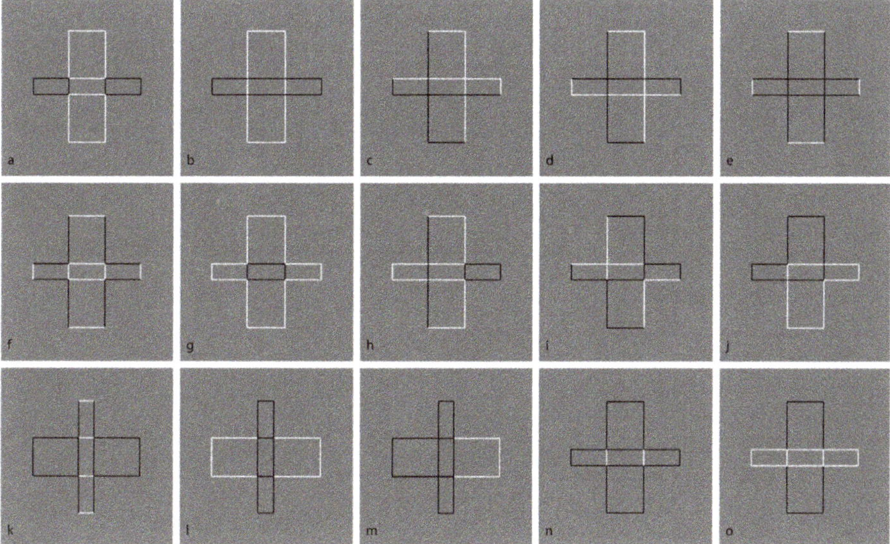

Figure 10. The same set of stimuli of Figure 7 has been redrawn by introducing Petter's effect according to which the larger rectangle is expected to be seen in front of the thinner one.

There is another, more interesting factor playing in the two counterparts. While in Figure 7a, all the elements are equal in size, in Figure 10a, they are dissimilar in pairs: The two horizontal rectangles differ in size from the vertical ones. This dissimilarity is another factor that favors the amplified effect of segregation in spite of the stronger Petter's effect that is expected to weaken this segregation. This result suggests that Petter's effect and, more generally, amodal completion operates after the grouping of the elements occurs on the basis of more primitive and general principles of grouping and figure–ground segregation. The same kind of argument can be used against the global results expected on the basis of simplicity/Prägnanz, unconscious inference, and Bayes' inference. All the remaining stimuli of Figure 10 obey to the same rationale.

In Figure 11, Petter's figure, as illustrated in Figure 9b, has been extended to the entire set of previous stimuli. In the basic Figure 9b, the similarity/dissimilarity between the two kinds of contours is a further principle useful to test the strength of the contrast polarity. More particularly, in addition to the grouping defined by good continuation and T-junctions, the contour similarity is now used in favor or against the similarity imparted by contrast polarity.

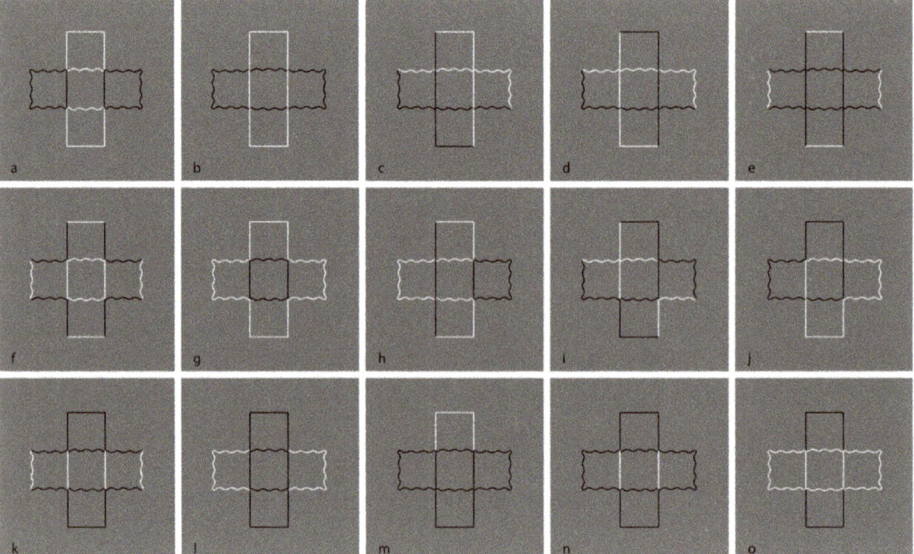

Figure 11. The same set of stimuli of Figure 7 has been redrawn by introducing Petter's effect according to which the vertical rectangle with straight boundaries is expected to be perceived in front of the one with undulated contours.

The descriptions of the stimuli were basically the same as their counterpart in Figure 7. The resulting effects, though very effective, were judged slightly weaker in terms of salience and figure–ground segregation.

The only difference between the two sets of stimuli relates to Figure 11d, where the convex–concave effect between the two figures is reversed. This was done to test the same kind of phenomenal attribute, namely, the shape which should be in front is concave and the one which should be seen behind is convex. Therefore, although the two figures are different, their visual meaning is the same.

3.4. Contrast Polarity and Tessellation

The figures studied in the previous sections mostly demonstrated the role of contrast polarity in weakening or breaking T-junctions, good continuation, and then amodal completion, both in classical and in Petter's conditions. There is a limiting case of amodal completion that effectively demonstrates the role of good continuations and of the related junctions. This is the tessellation. We have already seen in Figure 6 potential cases of tessellation made up of adjacent squares. They are like juxtaposed pavement tiles. Phenomenally, this is not the best condition to obtain a true tessellation, since the junctions are such that they let the good continuation flow without changing directions. Therefore, the result is a grid of vertical and horizontal contours, which, in Figure 6, group in two intersecting orthogonal rectangles.

A stronger case of tessellation uses hexagons instead of squares, as illustrated in Figure 12. Under these conditions, in terms of the good continuation principle, the bifurcation designed by the Y-junctions elicits two possible directions in perfect equilibrium. Therefore, the flow of good continuation is not stopped by any boundaries responsible for the amodal continuation but proceeds seamlessly. This kind of fluid dynamics description is a metaphorical way to make explicit the role of good continuation with this kind of junction.

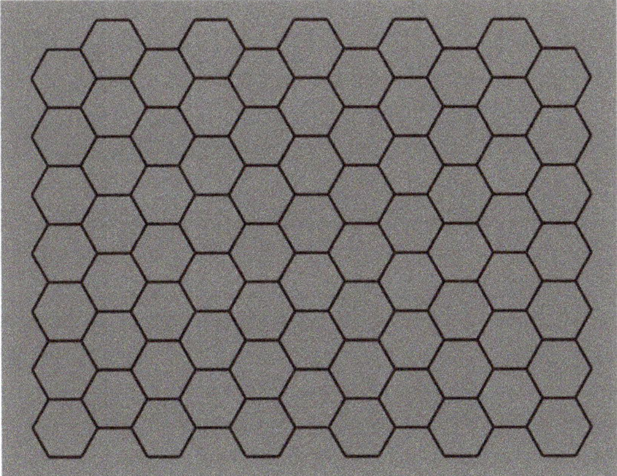

Figure 12. Phenomenal hexagonal tessellation.

Given the grouping/ungrouping effectiveness of contrast polarity, the main question to be answered here is the following: Can contrast polarity elicit amodal completion within a univocal condition of tessellation like the one illustrated in Figure 12?

In Figure 7o, we have seen that reversed contrast switched T-junctions to I-junction. Now, to answer the previous question, contrast polarity is supposed to be able to elicit amodal completion without changing the kind of junctions, i.e., to induce amodal completion with Y-junctions.

In Figure 13, the number of hexagonal cells has been reduced to only four to annul potential global and reference effects of the contest. Figure 13a can be considered the simplest case of amodal completion according to which, by turning one of the four hexagons into a white one, it induces a depth segregation and a change in the inner shape configuration of the two adjacent hexagons that continue amodally behind the white one. Therefore, the oblique sides of the two black hexagons converge amodally and meet in about the center of the white hexagon.

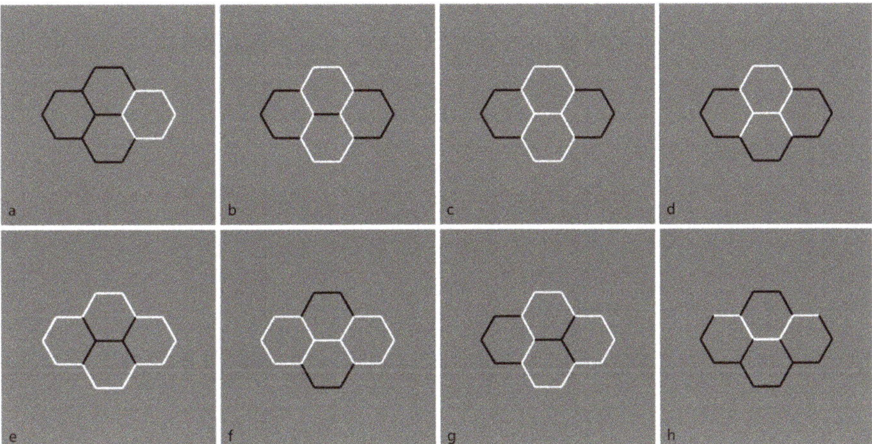

Figure 13. Different kinds of amodal completion induced by contrast polarity within hexagonal tessellations.

In Figure 13b, the white boundaries become a unique object and, given the unilateral belongingness of the boundaries, induce an amodal completion of the left and right black components that appear like a rectangle-like shape with pointed sides. An analogous amodal result was reported for Figure 13c.

The figure perceived amodally in Figure 13d is a black shape with three pointed sides. In Figure 13e, the white contours are perceived as the boundaries of a four-pointed shape and the inner black elements are its decoration or the visible parts of something larger seen through a window.

In Figure 13f, the upper and lower black contours complete amodally as a kind of rhombic shape behind the two white joined and filled hexagons. Some observers reported something equivalent but in different terms: a diamond face with white glasses.

Figure 13g shows a white shape with three pointed corners in front of some kind of hexagonal shape. It is worth noting that the black contours inside the white shape do not belong to the amodal hexagonal shape but to the white one.

Finally, in Figure 13h, the white contours separating the upper and lower black components are seen as belonging unilaterally to the lower hexagons; therefore, the upper black contours continue amodally behind them.

The role of contrast polarity can be expanded and generalized within a larger tessellation (as in Figure 12), thus creating the effective amodal organizations illustrated in Figure 14. Depth segregation and unilateral belongingness of the boundaries are imparted by the contrast polarity in two regions, one bulging closer to the observer and capturing the inner black contours and the other as an irregular window or as a hole showing the white net of contours in depth.

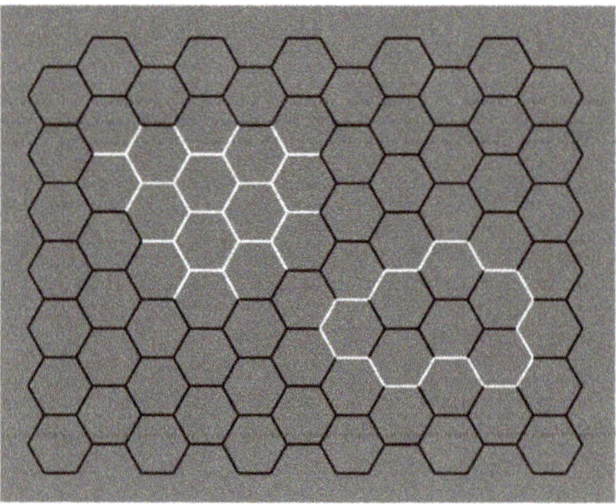

Figure 14. Two different kinds of amodal completion within hexagonal tessellations.

In Figure 15, two polygonal white shapes are perceived as joined together in some kind of annulus or alternately as two overlapping shapes, with the larger behind the smaller one. The depth segregation of the surrounding black hexagons is also clearly visible.

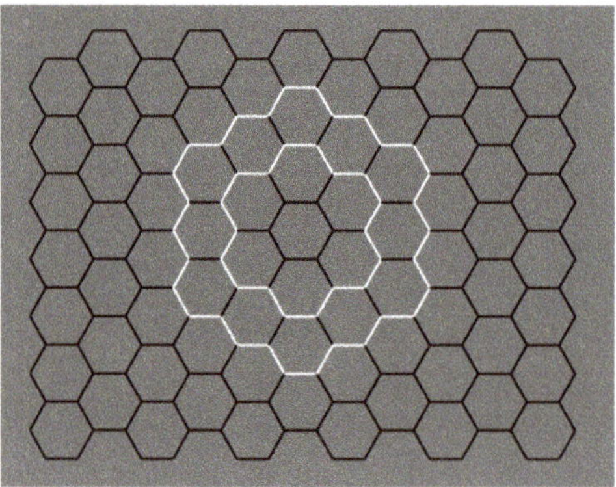

Figure 15. Two polygonal white shapes are perceived as joined together in some kind of annulus or, alternately, two overlapping shapes with the larger behind the smaller one.

A further simple but more effective condition is illustrated in Figure 16, where isolated alternated hexagons have been highlighted with white boundaries. Behind them, the amodal perceived texture is very different from the surrounding hexagons, although they should trigger the "unconscious inference" of a homogeneous hexagonal texture according to simplicity and Bayes' framework.

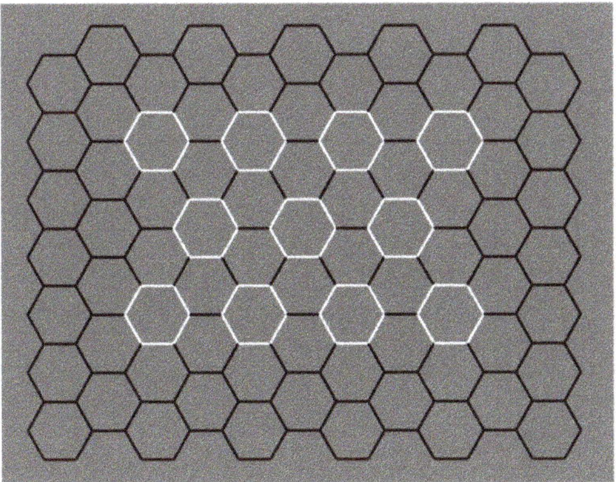

Figure 16. White hexagons partially occluding an amodal texture, different from the surrounding black hexagons.

It is worthwhile to focus on some phenomenal details within this stimulus. First of all, a deformation of the black polygons adjacent to the white ones is perceived according to their amodal continuation behind them. Second, the inner grey of the white hexagons appears different from the one of the black hexagons. This is likely due to contrast-assimilation effects, not further deepened here.

The results of this section support the important role of contrast polarity in perceptual organization. Its strength has been tested here against tessellation, which depends on good continuation and Y-junctions. By taking into consideration all these outcomes, the grouping and ungrouping strengths of contrast polarity could be phenomenally interpreted as a consequence of the border ownership that, in turn, is imparted by the visual salience and highlights effects due to the largest contrast among elements and, thus, by the similarity/dissimilarity among white and black elements. In terms of visual salience, the emergence of the components highlighted through the contrast polarity is, indeed, much stronger than the emergence due to the good continuation alone with its junctions. We suggest that it is exactly the popping out imparted by the reversed contrast that is mainly responsible for the described phenomena. Its salience effect is so strong it dominate other principles when it is pitted against them.

While these approaches are supposed to operate more appropriately at a global and holistic level of vision, explaining the whole percept beyond the single local components, the contrast polarity is a property of the stimulus and a visual attribute that operates locally, eliciting results that could be independent from any global scale and even paradoxical, with contradictory object hypothesis or without any likelihood, as shown in Figure 8 and as reported in the next section. The local highlighted elements of the contrast polarity also manifest global effects when they group together on the basis of the similarity/dissimilarity principle that segregates black and white components as belonging to different visual objects.

3.5. Contrast Polarity and I-Junctions

In this section, the I-junctions and the contrast polarity will be tested in limiting conditions that cannot be otherwise explained with the theories considered here. Before introducing the crucial cases, let us first go back to a known effect studied by Kanizsa [9,10,66].

He demonstrated that the shapes of the four partially occluded figures, as illustrated in Figure 17, are not determined by the Gestalt principle of symmetry. In spite of the symmetry and regularity principles suggested by their modal visible portions the amodal shapes perceived are instead influenced by the good continuation of the contours intersecting the occluding square. As a consequence, the amodal completions of the stimuli in Figure 17 appear as four asymmetrical shapes.

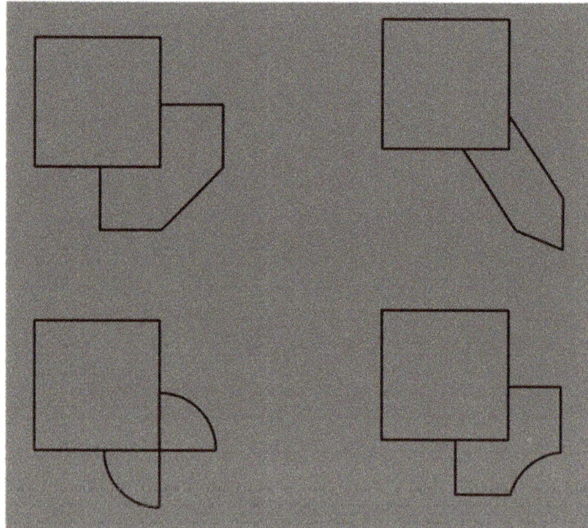

Figure 17. The figures, partially occluded by the square, complete amodally as asymmetrical shapes.

By introducing the contrast polarity as a new player, the occluder–occluded relationship and the amodal completion are reversed (see Figure 18). What previously have been partially hidden by the occluding square now appears as an occluder, and the square in the four conditions is seen partially occluded in its lower right side and exactly on the boundary contours. The I-junction effect is clearly perceived together with the black boundary of the square amodally continuing behind the white boundaries of the white emerging figures. This result occurs also when the luminance contrast among the components is inverted (not illustrated). Again, contrast polarity dominates over good continuation, T-junctions, and regularity.

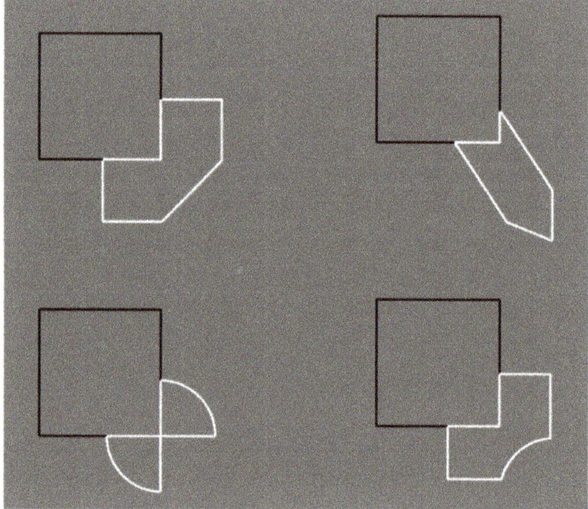

Figure 18. The contrast polarity reverses the amodal completion of Figure 17: The square is now perceived as partially occluded by the asymmetrical objects.

It could be noted that the third stimulus with two circular sectors with opposed corners due to contrast polarity turns back from irregular, as perceived in Figure 17, to regular in Figure 18. Kanizsa's argument against regularity and in favor of good continuation is reversed here.

Some might argue that, in Figure 18, the geometrical conditions are restored against the phenomenal ones. This is not totally true since, in Figure 18, amodal completion is still occurring but it is simply reversed in relation to the one perceived in Figure 17.

Within the likelihood and the avoidance-of-coincidences principles, these results are not expected, since the visual system is supposed to prevent interpretations elicited by coincidences. In all the critical conditions studied in this work, the coincidences are not discharged but highlighted.

The next step is to extend the I-junction effect to conditions where there are no junctions at all but just isolated close figures, like the four stars arranged to create a whole cross, as shown in Figure 19a. It should be noted that the perceived cross is just virtual, namely, a simple arrangement of figures. As a matter of fact, there is not any modal or amodal completion that can combine and unite the four stars that are perceived as such. Figure 19b shows how to join them by reversing the contrast of the four close sides near the center of the group of stars.

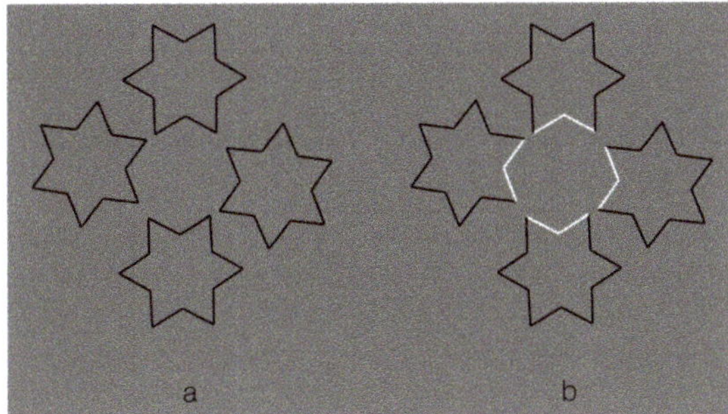

Figure 19. From four stars to a cross partially occluded by a white square-like shape.

Under these conditions, they are not stars anymore but, due to their amodal completion behind the white square shapes, they become a unity, a unique object more similar to a cross (not virtual) with its extreme parts similar to stars.

The following final step is a further extension of our rationale to a limiting case made up of a single shape, i.e., a closed contour, which, by definition, univocally and trivially should be assumed as a unique indivisible object or, at least, made up of sides and angles without any amodal completion that implicitly and explicitly requires two objects placed at different depths.

3.6. The Limiting Case of Amodal Completion

As our analysis proceeds, we demonstrate that the strength of the reversed contrast is so great in defining visual objects and amodal completion that it can be considered the optimal tool to test the appropriateness of the theories of which the main purpose is to answer the fundamental question "What is a perceptual object?".

Since amodal completion can be inverted in conditions where multiple elements play together to create different objects, the question is now the following: Can amodal completion be instilled along the boundary contours of a single shape?

This is indeed the limiting case useful to explore the perceptual principles involved in amodal completion and to test the maximum effectiveness of the contrast polarity. The paradigm of the limiting conditions is reminiscent of the one used in physics to explore the properties of elementary particles, and it is opposite to the holism embraced by Gestalt psychologists [45], mainly based on the assumption that the whole is greater than the sum of its parts, thus implying that the whole cannot be broken down.

Following this paradigm, a detrimental and disproving limiting case for both the simplicity and likelihood principles are the conditions illustrated in Figure 20, where the black boundaries of a regular eight-pointed star have been discontinued by reversing the contrast polarity in different portions of each figure. In terms of grouping, the closure principle is mainly responsible for the perception of the eight-pointed star. Also, the good continuation plays a role. Since there are no junctions, the flow of continuation is always good without bifurcations to be chosen or barriers to be amodally completed or continued.

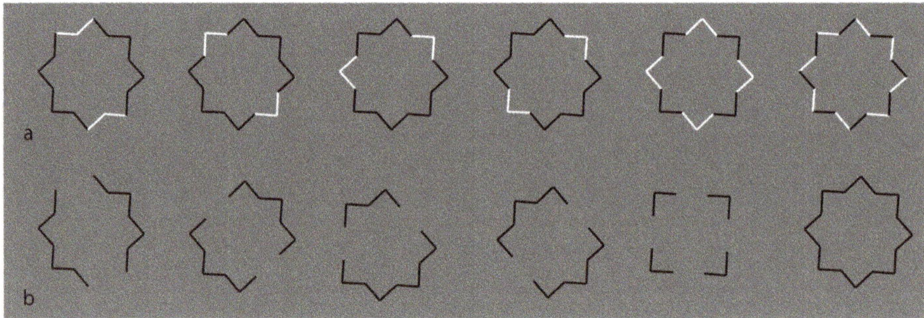

Figure 20. Contrast polarity breaks the unitariness of the stars and elicits amodal completion without junctions (cfr. (**a**) and (**b**) rows).

Phenomenally, the uniqueness and wholeness of each star illustrated appear weakened or broken down. Each shape of the first row partially splits into two black and white adjacent components that are seen as belonging to two different things. The two things do not show a traditional amodal completion but something that is anyway comparable to it. In fact, the most extensive black components more easily reveal the formation of a whole shape that, given the segregated white sides become "something else", do not appear as eight-pointed stars but as the shapes illustrated in Figure 20b. These shapes, shown here as fragments, tend to complete amodally behind the white sides. Although the removal of the white components improves the virtual grouping of the black segments, they appear more clearly as complete and full objects only in Figure 20a, where the closure and good continuation principles are fully functional. The missing sides of Figure 20a do not affect the whole shape but appear as gaps favoring the good continuation of the modal sides and the amodal wholeness [7,68–70]. Given the black and white alternation of the sides on the last condition of the first row, the last condition of the first row appears analogous to the star of the last condition of the second row.

Further examples are illustrated in Figure 21, where the reverse contrast breaks the oneness and the unitariness of the eight-pointed star and significantly influences its shape. New organizations emerge as follows: a concave polygonal shape rather than a star (Figure 21a); two rotated and perpendicular intersecting square shapes with illusory curved sides (Figure 21b); and irregular shapes different from the stars in Figure 21c–g. Figure 21h is the control.

Figure 21a,b displays something that is worth noticing, namely the emergence of the concavity and of the convexity within the star. As a matter of fact, the two figures could be perceived as stars. However, the first appeared as a sort of "eight-concave star" or "eight-indented star" while the second was more similar to a convex star when the subjects were forced to perceive a star. The first, more vivid, and immediate outcome is the one reported above.

The breaking of the uniqueness and unitariness is very effective even though good continuation, closure, and even Prägnanz together favor the grouping. The dissimilarity imparted by the reversed contrast splits the figure. The question is why are these irregular and parceled solutions preferred to a simpler result like "a star with black and white edges"? With this solution, both the whole shape and the local changes would have been preserved and saved. Even if this ideal solution is possible, it does not prevail.

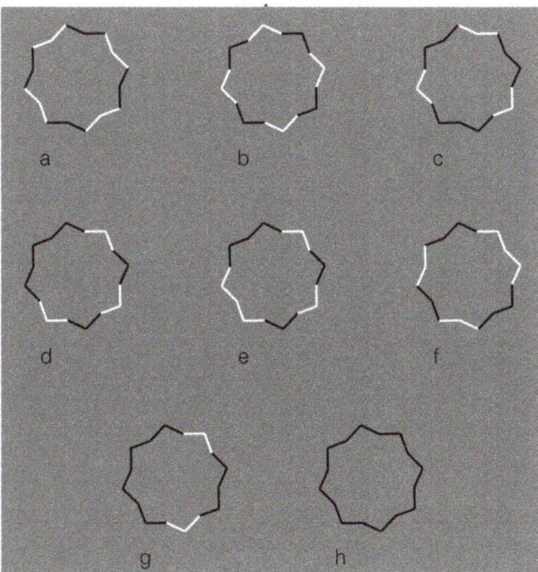

Figure 21. Contrast polarity breaks the unitariness of the stars and induces amodal completion without junctions.

3.7. The Limiting Case of the Limiting Case of Amodal Completion

The limiting case of amodal completion is very useful to make explicit the inner phenomenal dynamics. First of all, the contrast polarity can be considered like a barrier analogous but stronger than the one created by T-junctions. Most of our stimuli demonstrated this statement. This barrier is set up by the highest luminance contrast between the black and white poles on the gray background. This is the immediate corollary. This barrier is like a high step phenomenally stating "here starts something different, a new thing."

The result is the phenomenal salience of something highlighted or accentuated. As a matter of fact, the white elements placed on a gray background pop up with the greatest emphasis since they play with black elements placed on the other extreme of the contrast pole.

It follows that similar highlighted elements can group together on the basis of the similarity principle to elicit a new visual object, to which the boundaries unilaterally belonging. In other terms, the emphasized salient elements group or ungroup on the basis of their similarity or dissimilarity with the remaining elements, thus creating an occluder–occluding duality and, more generally, an effective object segregation and figure–ground organization.

This is a true sequential dynamics, according to which "the first is the phenomenal salience and, then, the similarity/dissimilarity". This can be proven through the final set of stimuli illustrated in Figure 22.

In the first column of Figure 22, the two black polygons are the controls: two octagons, on which the main vertical and horizontal axes are placed, respectively, along the corners (top) or on the sides (bottom). They are perceived as two different octagons with accentuated corners or sides. For a full description of these and related figures based on the accentuation principle, see Pinna [11,68,69].

The main predictions can be summarized as follows. The contrast polarity induces a phenomenal salience within the boundaries of the octagon. The salience can become (i) a full object through the similarity principle or (ii) an accent for the attributes of the shape location where the reversed contrast is placed. It becomes a full object if a sufficient number of similar components are present and then grouped. It appears as an accent when few elements or only one play(s) within a large set of elements with opposite contrast.

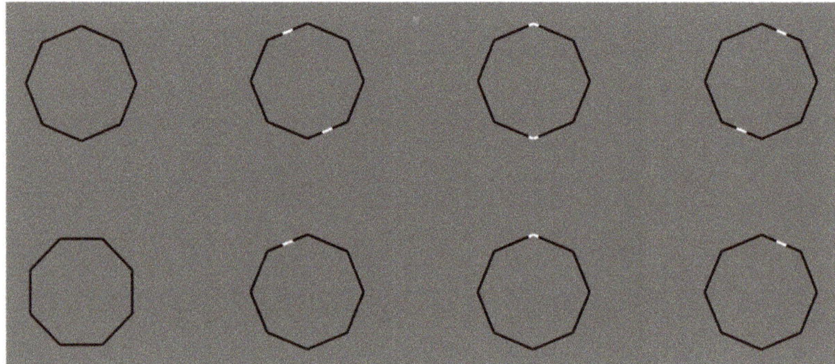

Figure 22. Octagons apparently different due to the accentuation imparted by the contrast polarity.

The creation of a full object has been already demonstrated in Figures 20 and 21. The accent emergence can be tested and seen in the other stimuli of Figure 22, where two (first row) or one (second row) polar accentuation(s) is/are placed on adjacent sides (second and fourth column) or on the top corner of the octagon (third column). All these figures are geometrical replicas of the first octagon on the top resting on one of its corners.

Phenomenally, when the polar accentuations are placed on the sides, the octagons are perceived as tilted (22.5 deg.) polygons, previously resting on one of their sides. They are perceived as more similar to the octagon at the bottom of the first column, although they are geometrically identical to the octagon on the top. Furthermore, they are perceived as tilted in opposite directions in relation to the vertical axis: the first tilted on the left and the second tilted on the right.

When the accentuation is placed on the corners, the octagon is seen like the octagon at the top of the first column. Only one accent (second row) is sufficient to highlight the attributes of the figure where the accent occurs. We called these attributes sidedness and pointedness [58,68,82,106].

Within this figure, we leave out the letters, which are useful to distinguish and name each stimulus, since the letters can also play like accents although less strongly than the contrast polarity [71,107,108].

When the observers were suggested to perceive and report about the segregation or belongingness of the white spots, the most frequent answers were that they can be seen both as part of the polygon or as something different, independent, attached, and extra and as an intruder located above the boundary contours of the polygons.

These outcomes suggest that the white spots behave like true emerging objects placed in front of an amodal black polygon. If this is true, the conditions illustrated in Figure 22 can be considered the limiting case of the limiting case of amodal completion.

4. Discussion

In this work, we demonstrated unique and relevant visual properties imparted by contrast polarity in perceptual organization and, more particularly, in eliciting amodal completion that is one of the most common and interesting visual phenomena and compelling issues of vision science. Amodal completion is the vivid completion of a single continuous object of the visible parts of an occluded shape despite portions of its boundary contours not actually being seen. Psychophysical data have demonstrated that T-junction, good continuation, and closure are the main principles involved [1,2,7,14–18].

We discussed the most relevant explanations of amodal completion based on Helmholtz's likelihood principle [28] and Gregory's "unconscious inference" [29,30]. According to these theories, the amodal object is similar to a perceptual hypothesis postulated to explain the unlikely gaps within the stimulus pattern and the one that most likely produces the sensory stimulation. Along the same theoretical line, Rock [31,32] proposed the so-called avoidance-of-coincidences principle, stating that the visual system tends to prevent interpretations elicited by coincidences [8,33–36]. More recently these approaches have been reconsidered in terms of probabilistic Bayesian inference [37–43], applied successfully in many classical amodal conditions [42,43]. Bayes' theorem computes a probabilistic decision-making aimed at choosing the amodal outcome as the result of the convolution between the prior p(H), modelling the constraint, and prior assumptions on the structure of the environment necessary to solve underdetermination and the conditional p(D|H) that models optics and the projection on the retina.

Another approach consistent with the previous ones and largely used to explain amodal completion is based on the simplicity–Prägnanz principle of Gestalt psychologists, according to which the visual system is aimed at finding the simplest and most stable organization consistent with the sensory inputs.

All these approaches focus attention mostly on the shape that amodally completes the visible fragments; therefore, they assume amodal completion as the cause of the amodal shape formation. The main interest of this perspective is to explain how the occluded object is completed, what is the amodal shape, and how contours of partially visible fragments are relatable behind an occluder.

Different from these approaches, we adopted the complementary perspective, assuming amodal completion not as the cause but as the resulting effect. The related questions we addressed through our stimuli are the following: "What is the role of shape formation and perceptual organization in inducing amodal completion? What are the perceptual conditions that elicit the segregation of occluded and occluding objects and, finally, amodal completion? What is the role of the local contours, junctions and termination attributes in eliciting the phenomenon of amodal completion?"

Within this perspective, amodal completion has been considered here as a visual phenomenon not as a process, i.e., the final outcome of perceptual processes and grouping principles. Moreover, the contrast polarity with its related similarity/dissimilarity outcomes has been used as the main grouping and ungrouping attribute to explore amodal completion as a visual phenomenon elicited by good continuation, T-junctions, and regularity.

The stimuli were designed as *instantiae crucis* (crucial instances) and studied through the experimental phenomenology, according to the general methods used by Gestalt Psychologists. Together with traditional stimulus configurations, we introduced novel patterns of stimuli, which have been reduced more and more to extreme limiting conditions.

Through our stimuli, contrast polarity has been demonstrated to be effective in inducing amodal completion in conditions where, on the basis of the known principles and of the previous theoretical approaches, it is not expected and vice versa: Amodal completion was annulled or disrupted in the patterns of stimuli where it was supposed to be effective. More in detail, contrast polarity has been able to elicit/disrupt amodal completion when pitted against or in favor of the following conditions reduced to limiting cases: (i) classical patterns (Figures 7 and 8); (ii) Petter's effect and Petter's rule (Figures 9–11); (iii) tessellation with T-junctions replaced by Y-junctions (Figures 12–16); (iv) group of isolated figures arranged in a cross (Figures 17–19); and (v) a single shape (Figures 20–22).

The results demonstrated the domination of the contrast polarity against good continuation, T-junctions, and regularity. Moreover, the limiting conditions explored revealed a new kind of junction next to the T- and Y-junctions, respectively responsible for amodal completion and tessellation. We called them I-junctions. They elicited the amodal continuation of contours behind a contour with the same orientation.

Contrast polarity was shown to operate locally, eliciting results that could be independent from any global scale and that could also be paradoxical. These results weaken and challenge theoretical approaches based on notions like oneness, unitariness, symmetry, regularity, simplicity, likelihood, priors, constraints, and past knowledge. Therefore, Helmholtz's likelihood principle, simplicity/Prägnanz, and Bayes' inference were clearly questioned since they are supposed to operate especially at a global and holistic level of vision.

An alternative explanation of the specific outcomes, particularly related to the limiting conditions of amodal completion, could be based on the phenomenal dynamics made explicit by contrast polarity. First of all, the contrast polarity was perceived like a barrier analogous but stronger than the one created by T-junctions. Most of our stimuli demonstrated this general statement. From this, as a corollary, it follows that this barrier is raised by the highest luminance contrast between the black and white poles on the gray background. This barrier phenomenally represents the starting point for a new emergent "thing", a new attribute, or a new object.

In other terms, the phenomenal salience, elicited by the highest luminance contrast going from black to white on a gray background, triggers a process of object segregation and its related dynamics: unilateral belongingness of the boundaries and similarity/dissimilarity principles.

The dominance of contrast polarity over good continuation and T-junctions is related to its stronger phenomenal salience and highlight effect. The same argument can account for the emergence of amodal continuation on I-junction, groups of isolated figures, and single shapes. Moreover, the imparted salience can disrupt, both locally and globally, arrangements of figures or can alternately rearrange the elements according to their similarity/dissimilarity. The highlighting strength of contrast polarity determines even the grouping effectiveness against the global and holistic rules and factors expected by Helmholtz's likelihood principle, simplicity/Prägnanz, and Bayes' inference.

In favor of the basic and essential role of the phenomenal salience, we can invoke deceiving strategies used in nature by most living organisms. Flowers, birds, and fishes use colors and contrast polarity to attract, reject, show, and hide: to show some parts more clearly than others; to show something that would be otherwise invisible; to show parts that are not natural parts; to show fragments; to show in order to hide; to show not to show; to show to break and split; to show to separate; to show to multiply; to show the oneness; to show some elements, some more or less important elements; to show something and not to show something else; and to show some parts and not to show the whole [85,109]. In short, the phenomenal salience strongly improves the biological fitness of living organisms and, therefore, the capability of an individual of a certain genotype to reproduce and, thus, to propagate an individual's genes within the genes of the next generation.

The phenomenal salience is a basic requirement also in human beings, in the way we dress, invent fashion and design, in the way we use the maquillage, and in the existence of the maquillage itself [110].

The strength of phenomenal salience imparted by contrast polarity enables the full independence from local or global organizations and top-down or bottom-up dynamics. It eludes all these categories since it can play in favor or against each of them, as demonstrated in our stimuli. It represents a true challenge for the theories discussed here, which cannot easily incorporate it (e.g., as a prior) without losing explanatory power somewhere else. Inside these arguments, it is important to underline that phenomenal salience is a perceptual attribute not restricted to contrast polarity, but it can also be triggered by color, shape, motion and every other visual property [109]. Among them, contrast polarity is one of the most powerful.

Given the significance of this attribute, further experimental studies based on phenomenological, psychophysical, and neurophysiological techniques are required to measure the strength of the phenomenal salience imparted by contrast polarity against other attributes involved in amodal completion. Further studies can shed light on the role of contrast polarity as a general tool useful in testing the range of scientific effectiveness of visual theories, approaches, and models.

Finally, our phenomenological results suggest several extensions and implications for vividness, imagery, and consciousness, shortly described below.

Implications for Vividness, Imagery, and Consciousness

There are several phenomenological elements that might turn the scientific attention from amodal completion to the notion of vividness, considered as a property expressing the self-rated degree of richness, details, resolution, and clarity of a mental image, as compared to the experience of actual perception [111] (D'Angiulli & Reeves, 2007). Although this simple definition is supported by a plethora of correlations, studied and measured in literature [112–121], the function and underlying processes of vividness are still sources of deep scientific challenges and theoretical controversies [111,122–124]. Without getting too deep into detail, we suggest that our work might make a contribution by placing new clues and ideas on the debate about the role of vividness in cognitive neuroscience and of its meaning as a phenomenological component of consciousness.

To show how amodal completion could be related to the notion of vividness, we go back to the simple definition of amodal completion reported here: Amodal completion is the vivid outcome of a complete object unity, i.e., the vivid completion as a single continuous object of the visible parts of an occluded shape despite portions of its boundary contours not actually being seen. For our purposes, two interesting terms of this description that are worth highlighting are "vivid" and "completion". Phenomenally, they belong to close domains although different: "Vivid" is a clear visual outcome under our conditions, and "completion" is in between vision and imagery. This is due to the fact that, even though the sensory experience of completeness and unity is perceptual, the portions of boundary contours actually seen are not directly visible, namely amodal. This distinction becomes more salient under complex, simple, or uncertain conditions (e.g., some of the stimuli shown in the previous sections), where completion elicits different and alternative solutions and mental image formations. Moreover, as a supporting common result, naive subjects spontaneously assume amodal completion as a sort of mental construction (completion) of the invisible part of an object due to past experience, imagery, or some kind of memory association. This kind of ingenuous theory emerges very often at the end of an experiment, testing amodal completion, when subjects ask for its meaning.

Related to the term "completion" is "amodal", commonly accepted by the scientific community and defined as the vivid experience of completeness without seeing the occluded part of the object. The expression "without seeing" suggests that something that cannot be seen is clearly perceived. This is definitely in between something perceived and something that is not perceived. The need to introduce the term "amodal" is aimed at explaining this uncertain and twofold meanings. Again, the best conditions to perceive the in-between placement of the term "amodal" are specific experimental tasks or ambiguous-complex stimuli (e.g., Figures 6 and 7), requiring time to be clearly seen and described. Under these circumstances, the amodal completion of possible emerging percepts manifests different degree of vividness.

Indeed, the phenomenological task adopted here can be considered a very functional tool to explore the complexity of these terms, which are, in turn, interesting phenomena to be explored. Moreover, testing simple stimuli, like the previous ones, where grouping principles are pitted one against the other, is useful to explore the no man's land where vision and imagery meet. Although in everyday life, we perceive effortless complete objects, under unfamiliar, crowded, or camouflaged conditions, the terms "vivid", "completion", and "amodal" can easily assume the twofold and in-between meanings described.

A further element is the notion of amodal completion considered as a visual phenomenon, i.e., the final outcome of perceptual processes and grouping principles. This entails that object formation comes first before amodal completion. In short, the priority is to perceive objects, while amodal completion is usually a hidden or a secondary phenomenon. This kind of phenomenal organization clearly improves the biological fitness of living organisms and imparts explicit psychological advantages. Indeed, the need to perceive or create a mental image of an occluded object is compelling since objects are true invariants required in the foreground, while amodal completion can vary widely and can therefore be placed in the background.

It is not a coincidence that, to perceive amodal completion as a phenomenon, naive people need to be trained; otherwise, it remains in the background or totally invisible. This is also what occurred in the history of vision science. Despite how prevalent and important is this phenomenon, it went unnoticed by scholars for a long time.

More generally, as many other object properties and meanings, amodal completion does not pop up perceptually with the same salience and vividness as others, but it is less prominent and placed at the lower steps of the so-called "gradient of visibility" [59,71].

A final element, even more clearly related to the notion of vividness, is the phenomenal salience, elicited by the reversed contrast. As demonstrated, the contrast polarity can easily play against or in favor of all the objects potentially included in Figure 6, by highlighting one or the other and, conversely, by hiding and making the others invisible. Again, this means that vision always plays under these conditions with modal and amodal perception along the higher and lower steps of the gradient of visibility and consciousness.

All these clues could be useful to explore the role of vividness acting as an index of availability of sensory information and traces and playing a basic phenomenological role in understanding the access to percepts and memories.

Finally, they suggest that the phenomenal salience imparted by reversed contrast and the gradient of visibility might be, as well as vividness, significant tools to isomorphically define, measure, and express the grades of the conscious experience that pops up from the brain's complexity of sensory information in perception. In other terms, they express variations and different states of inner first-person experience of the input and the environmental input itself.

5. Conclusions

In conclusion, the vividness of amodal completion under the conditions here studied, together with the phenomenal salience imparted by the reversed contrast and the related gradient of visibility, can be considered phenomenological features of primary sensory consciousness, thus supporting the hypothesis that consciousness is a graded phenomenon.

Author Contributions: Conceptualization, B.P.; methodology, L.C.

Acknowledgments: This work is supported by Fondazione Banco di Sardegna (finanziato a valere sulle risorse del bando Fondazione di Sardegna—Annualità 2015).

Conflicts of Interest: The authors declare no conflict of interest. The funders had no role in the design of the study; in the collection, analyses, or interpretation of the data; in the writing of the manuscript; or in the decision to publish the results.

References

1. Ekroll, V.; Sayim, B.; Wagemans, J. Against better knowledge: The magical force of amodal volume completion. *i-Perception* **2013**, *4*, 511–515. [CrossRef] [PubMed]
2. Ekroll, V.; Sayim, B.; Van Der Hallen, R.; Wagemans, J. Illusory Visual Completion of an Object's Invisible Backside Can Make Your Finger Feel Shorter. *Curr. Boil.* **2016**, *26*, 1029–1033. [CrossRef] [PubMed]
3. Michotte, A. *The Perception of Causality*; Basic Books: New York, NY, USA, 1946/1963.
4. Michotte, A.; Burke, L. *Une Nouvelle Énigme de la Psychologie de la Perception: le "Donnée Amodal" dans L'experience Sensorielle. Actes du XIII Congrés Internationale de Psychologie*; Publications Universitaires: Louvain, Belgium, 1962; pp. 347–371.
5. Michotte, A.; Thinès, G.; Crabbé, G. *Les Compléments Amodaux des Structure Perceptives*; Erlbaum: Hillsdale, NJ, USA, 1991; pp. 67–140.
6. Pinna, B. Illusory contours and neon color spreading reconsidered in the light of Petter's rule, 261–284. In *Information in Perception*; Albertazzi, L., van Tonder, G., Vishwanath, D., Eds.; MIT Press: Cambridge, MA, USA, 2011.
7. Pinna, B. The role of amodal completion in shape formation: Some new shape illusions. *Perception* **2013**, *41*, 1336–1354. [CrossRef] [PubMed]
8. Pinna, B.; Ehrenstein, W.; Spillmann, L. Illusory contours and surfaces without perceptual completion and depth segregation. *Vis. Res.* **2004**, *44*, 1851–1855. [CrossRef] [PubMed]
9. Kanizsa, G. *Organization in Vision*; Praeger: New York, NY, USA, 1979.
10. Kanizsa, G. Seeing and thinking. *Acta Psychol.* **1985**, *59*, 23–33. [CrossRef]

11. Pinna, B. A new perceptual problem: The amodal completion of color. *Vis. Neurosci.* **2008**, *25*, 415–422. [CrossRef]
12. Singh, M. Modal and Amodal Completion Generate Different Shapes. *Psychol. Sci.* **2004**, *15*, 454–459. [CrossRef]
13. Singh, M.; Hoffman, D.D.; Albert, M.K. Contour completion and relative depth: Petter's rule and support ratio. *Psychol. Sci.* **1999**, *10*, 423–428. [CrossRef]
14. Boselie, F.; Wouterlood, D. A good continuation model of some occlusion phenomena. *Psychol. Res.* **1992**, *54*, 267–277.
15. Kanizsa, G.; Gerbino, W. Amodal completion: Seeing or thinking. In *Organization and Representation in Perception*; Beck, J., Ed.; LEA: Hillsdale, NJ, USA, 1982; pp. 167–190.
16. Kellman, P.J.; Shipley, T.F. A theory of visual interpolation in object perception. *Cogn. Psychol.* **1991**, *23*, 141–221. [CrossRef]
17. Palmer, S.E. *Vision Science: Photons to Phenomenology*; MIT Press: Cambridge, MA, USA, 1999.
18. Sekuler, A.B.; Flynn, C.; Palmer, S.E. Local and Global Processes in Visual Completion. *Psychol. Sci.* **1994**, *5*, 260–267. [CrossRef]
19. Wertheimer, M. Über das Denken der Naturvölker. *Zeitschrift für Psychologie* **1912**, *60*, 321–378.
20. Wertheimer, M. Untersuchungen über das Sehen von Bewegung. *Zeitschrift für Psychologie* **1912**, *61*, 161–265.
21. Gilbert, C.D. Dynamic properties of adult visual cortex. In *The Cognitive Neuroscience*; Gazzaniga, M.S., Ed.; MIT Press: Cambridge, MA, USA, 1995; pp. 73–90.
22. Ts'o, D.Y.; Roe, A.W. Functional compartments in visual cortex: Segregation and interaction. In *The Cognitive Neuroscience*; Gazzaniga, M.S., Ed.; MIT Press: Cambridge, MA, USA, 1995; pp. 325–337.
23. Gazzaniga, M.S. Concurrent processing in the primate visual cortex. In *The Cognitive Neuroscience*; Gazzaniga, M.S. MIT Press: Cambridge, MA, USA, 1995; pp. 383–400.
24. Von der Heydt, R.; Peterhans, E. Mechanisms of contour perception in monkey visual cortex: I. Lines of pattern discontinuity. *J. Neurosci.* **1989**, *9*, 1731–1748. [CrossRef] [PubMed]
25. Anstis, S. Visual filling-in. *Curr. Biol.* **2010**, *20*, R664. [CrossRef] [PubMed]
26. Komatsu, H. The neural mechanisms of perceptual filling-in. *Nat. Rev. Neurosci.* **2006**, *7*, 220–231. [CrossRef] [PubMed]
27. Murakami, I. *Perceptual filling-in. Encyclopedia of Neuroscience*; Springer: Berlin, Germany, 2008.
28. Von Helmholtz, H. *Handbuch der Physiologischen Optik*; Leopold Voss: Leipzig, Germany, 1867.
29. Gregory, R. Cognitive contours. *Nature* **1972**, *238*, 51–52. [CrossRef]
30. Gregory, R.L. Perceptions as hypotheses. *Philosoph. Trans. Roy. Soc. London* **1980**, *B290*, 181–197. [CrossRef]
31. Rock, I. *The Logic of Perception*; MIT Press: Cambridge, MA, USA, 1983.
32. Rock, I. Perception and knowledge. *Acta Psychologica.* **1985**, *59*, 3–22. [CrossRef]
33. Biederman, I. Recognition-by-components: A theory of human image understanding. *Psychol. Rev.* **1987**, *94*, 115–117. [CrossRef]
34. Hoffman, D.D. *Visual Intelligence*; Norton: New York, NY, USA, 1998.
35. Binford, T. Inferring surfaces from images. *Artif. Intel.* **1981**, *17*, 205–244. [CrossRef]
36. Witkin, A.P.; Tenenbaum, J.M. On the role of structure in vision. In *Human and Machine Vision*; Beck, J., Hope, B., Rosenfeld, A., Eds.; Academic Press: New York, NY, USA, 1983; pp. 481–543.
37. Bülthoff, H.H.; Yuille, A.L. Bayesian models for seeing shapes and depth. *Comm. Theor. Biol.* **1991**, *2*, 283–314.
38. Feldman, J. Bayesian contour integration. *Percep. Psychoph.* **2001**, *63*, 1171–1182. [CrossRef]
39. Feldman, J. Formation of visual "objects" in the early computation of spatial relations. *Percept. Psychophys.* **2007**, *69*, 816–827. [CrossRef] [PubMed]
40. Feldman, J.; Singh, M. Bayesian estimation of the shape skeleton. *Proc. Natl. Acad. Sci.* **2006**, *103*, 18014–18019. [CrossRef] [PubMed]
41. Kersten, D.; Mamassian, P.; Yuille, A. Object Perception as Bayesian Inference. *Annu. Rev. Psychol.* **2004**, *55*, 271–304. [CrossRef] [PubMed]
42. Knill, D.C.; Richards, W. (Eds.) *Perception as Bayesian Inference*; Cambridge University Press: Cambridge, UK, 1996.
43. Simoncelli, E.P.; Adelson, E.H.; Weiss, Y. Motion illusions as optimal percepts. *Nat. Neurosci.* **2002**, *5*, 598–604.
44. Köhler, W. *Die Physischen Gestalten in Ruhe und im Stationären Zustand*; Springer Science and Business Media LLC: Berlin, Germany, 1920.

45. Koffka, K. *Principles of Gestalt Psychology*; Routledge & Kegan Paul: London, UK, 1935.
46. Hatfield, G.; Epstein, W. The status of the minimum principle in the theoretical analysis of visual perception. *Psychol. Bull.* **1985**, *97*, 155–186. [CrossRef]
47. Leeuwenberg, E.L.J. *Structural information of visual patterns: An efficient coding system in perception*; Mouton: The Hague, The Netherlands, 1968.
48. Leeuwenberg, E.L. Quantitative specification of information in sequential patterns. *Psychol. Rev.* **1969**, *76*, 216–220. [CrossRef]
49. Leeuwenberg, E.L.J. A Perceptual Coding Language for Visual and Auditory Patterns. *Am. J. Psychol.* **1971**, *84*, 307. [CrossRef]
50. Leeuwenberg, E.L.; Boselie, F. Against the likelihood principle in visual form perception. *Psychol. Rev.* **1988**, *95*, 485–491. [CrossRef] [PubMed]
51. Mach, E. *The Analysis of Sensations and the Relation of the Physical to the Psychical*; Dover: New York, NY, USA, 1959.
52. Boselie, F. Local versus global minima in visual pattern completion. *Percept. Psychophys.* **1988**, *43*, 431–445. [CrossRef] [PubMed]
53. Boselie, F. Local and Global Factors in Visual Occlusion. *Percept.* **1994**, *23*, 517–528. [CrossRef] [PubMed]
54. Boselie, F. The Golden Section and the Shape of Objects. *Empir. Stud. Arts* **1997**, *15*, 131–141. [CrossRef]
55. Rubin, E. *Synsoplevede Figurer*; Gyldendalske Boghandel: Copenhagen, Denmark, 1915.
56. Rubin, E. *Visuell wahrgenommene Figuren*; Gyldendalske Boghandel: Kobenhavn, Denmark, 1921.
57. Nakayama, K.; Shimojo, S. Toward a Neural Understanding of Visual Surface Representation. *Cold Spring Harb. Symp. Quant. Boil.* **1990**, *55*, 911–924. [CrossRef]
58. Pinna, B. New Gestalt principles of perceptual organization: An extension from grouping to shape and meaning. *Gestalt Theory* **2010**, *32*, 1–67.
59. Pinna, B. What Comes Before Psychophysics? The Problem of 'What We Perceive' and the Phenomenological Exploration of New Effects. *See. Perce.* **2010**, *23*, 463–481. [CrossRef]
60. Pinna, B. What is a perceptual object? Beyond the Gestalt theory of perceptual organization. In *Beyond Gestalt: Neural Foundations of Visual Perception*; Geremek, A., Greenlee, M., Magnussen, S., Eds.; Taylor & Francis: New York, NY, USA, 2013.
61. Pinna, B. Illusion and illusoriness: New perceptual issues and new phenomena. In *Handbook of Experimental Phenomenology*; Albertazzi, L., Ed.; Blackwell-Wiley: Hoboken, NJ, USA, 2013.
62. Singh, M. Visual representation of contour geometry. In *Handbook of Perceptual Organization*; Wagemans, J., Ed.; Oxford Library of Psychology: Oxford, UK, 2013.
63. Singh, M.; Fulvio, J.M. Visual extrapolation of contour geometry. *Proc. Natl. Acad. Sci. USA* **2005**, *102*, 939–944. [CrossRef]
64. Singh, M.; Hoffman, D.D. Part-Based Representations of Visual Shape and Implications for Visual Cognition. In *Time, Internal Clocks and Movement*; Elsevier BV: Amsterdam, The Netherlands, 2001; Volume 130, pp. 401–459.
65. Kanizsa, G. Schrumpfung von visuellen Feldern bei amodaler Ergänzung. *Studia Psychologica* **1972**, *14*, 208–210.
66. Kanizsa, G. Amodal completion and phenomenal shrinkage of surfaces in the visual field. *Ital. J. Psychol.* **1975**, *2*, 187–195.
67. Kanizsa, G.; Luccio, R. *Espansione di superficie da completamento amodale. Reports from the Institute of Psychology*; University of Trieste: Trieste, Italy, 1978.
68. Pinna, B. What is the meaning of shape? *Gestalt Theory* **2012**, *33*, 383–422.
69. Pinna, B. Perceptual Organization of Shape, Color, Shade, and Lighting in Visual and Pictorial Objects. *i-Perception* **2012**, *3*, 257–281. [CrossRef] [PubMed]
70. Pinna, B. The Place of Meaning in Perception—Introduction. *Gestalt Theory* **2012**, *33*, 3221–3244.
71. Pinna, B. Directional organization and shape formation: new illusions and Helmholtz's Square. *Front. Hum. Neurosci.* **2015**, *9*, 92. [CrossRef] [PubMed]
72. Albertazzi, L. *Philosophical Background: Phenomenology*; Oxford University Press: Oxford, UK, 2015.
73. Koenderink, J.J. *Methodological Background: Experimental Phenomenology*; Oxford University Press: Oxford, UK, 2015.

74. Pinna, B.; Deiana, K. Material properties from contours: New insights on object perception. *Vis. Res.* **2015**, *115*, 280–301. [CrossRef] [PubMed]
75. Verstegen, I. Mona Lisa's smile. The place of experimental phenomenology within gestalt theory. *Gestalt Theory* **2005**, *27*, 91–106.
76. Duhem, P. *La Théorie Physique: Son Object et sa Structure*; Chevalier et Rivière: Paris, France, 1914.
77. Popper, K.R. *Conjectures and Refutations*; Routledge and Kegan Paul: London, UK, 1969.
78. Popper, K.R. *Poscript to the Logic of Scientific Discovery, Volume 1: Realism and the Aim of Science*; Hutchinson: London, UK, 1983.
79. Köhler, W. *The Place of Value in a World of Facts*; Liveright Publishing Corporation: New York, NY, USA, 1938.
80. Köhler, W. *Gestalt Psychology*; Liveright: New York, NY, USA, 1947.
81. Metzger, W. *Psychologie*; Steinkopff Verlag: Darmstadt, Germany, 1963.
82. Pinna, B.; Koenderink, J.; van Doorn, A. The phenomenology of the invisible: From visual syntax to "shape from shapes". *Philosophia Scientae* **2015**, *19*, 5–29. [CrossRef]
83. Spillmann, L.; Ehrenstein, W.H. Gestalt factors in the visual neurosciences. In *Visual Neurosciences*; Chalupa, L., Werner, J.S., Eds.; MIT Press: Cambridge, MA, USA, 2004; pp. 1573–1589.
84. Metzger, W. *Gesetze des Sehens*; Kramer: Krankfurt-am-Main, Germany, 1975.
85. Reeves, A.; Pinna, B. Lighting, backlighting and watercolor illusions and the laws of figurality. *Spat. Vis.* **2006**, *19*, 341–373. [CrossRef]
86. Petter, G. Nuove ricerche sperimentali sulla totalizzazione percettiva. *Rivista di Psicologia, L* **1956**, *50*, 213–227.
87. Shipley, T.F.; Kellman, P.J. Perception of partly occluded objects and illusory figures: Evidence for an identity hypothesis. *J. Exp. Psychol. Hum. Percept. Perform.* **1992**, *18*, 106–120. [CrossRef]
88. Shipley, T.F.; Kellman, P.J. Spatiotemporal boundary formation: Boundary, form, and motion perception from transformations of surface elements. *J. Exp. Psychol. Gen.* **1994**, *123*, 3–20. [CrossRef] [PubMed]
89. Koenderink, J.J.; van Doorn, A.J. Illuminance texture due to surface mesostructure. *J. Opt. Soc. Am. A* **1996**, *13*, 452. [CrossRef]
90. Koenderink, J.J.; van Doorn, A.J. Local structure of Gaussian texture. *IEEE Trans Inf. Syst.* **2003**, *86*, 1165–1171.
91. Koenderink, J.J.; van Doorn, A.J.; Kappers, A.M.L.; Pas, S.F.T.; Pont, S.C. Illumination direction from texture shading. *J. Opt. Soc. Am. A* **2003**, *20*, 987. [CrossRef]
92. Koenderink, J.J.; van Doorn, A.J.; Pont, S.C. Light direction from shad(ow)ed random Gaussian surfaces. *Perception* **2004**, *33*, 1405–1420. [CrossRef] [PubMed]
93. Koenderink, J.J.; Pont, S.C.; van Doorn, A.J.; Kappers, A.M.L.; Todd, J.T. The visual light field. *Perception* **2007**, *36*, 1595–1610. [CrossRef] [PubMed]
94. Koenderink, J.J.; van Doorn, A.J.; Pont, S.C. Shading, a view from the inside. *See. Perceiv.* **2012**, *25*, 303–338.
95. Koenderink, J.J.; van Doorn, A.J.; Wagemans, J. SFS? Not likely! *i-Perception* **2013**, *4*, 299–302. [CrossRef] [PubMed]
96. Koenderink, J.; van Doorn, A.; Pinna, B. Psychogenesis of Gestalt. *Gestalt Theo.* **2015**, *37*, 287–304.
97. Koenderink, J.; van Doorn, A.; Pinna, B.; Wagemans, J. Boundaries, Transitions and Passages. *Art Percept.* **2016**, *4*, 185–204. [CrossRef]
98. Mamassian, P.; Knill, D.C.; Kersten, D. The perception of cast shadows. *Trends Cogn. Sci.* **1998**, *2*, 228–295. [CrossRef]
99. Mamassian, P.; Goutcher, R. Prior knowledge on the illumination position. *Cognition* **2001**, *81*, B1–B9. [CrossRef]
100. Mamassian, P.; Landy, M.S.; Maloney, L.T. Bayesian modelling of visual perception. In *Probabilistic Models of the Brain: Perception and Neural Function*; Rao, R., Olshausen, B., Lewicki, M., Eds.; MIT Press: Cambridge, MA, USA, 2002; pp. 13–36.
101. Kolmogorov, A.N. Three approaches to the quantitative definition of information. *Problems Infor. Trans.* **1965**, *1*, 1–7. [CrossRef]
102. Wertheimer, M. Untersuchungen zur Lehre von der Gestalt. *Psychol. Res.* **1922**, *1*, 47–58. [CrossRef]
103. Wertheimer, M. Untersuchungen zur Lehre von der Gestalt. II. *Psychol. Res.* **1923**, *4*, 301–350. [CrossRef]
104. Stoner, G.R.; Albright, T.D. Image segmentation cues in motion processing: Implications for modularity in vision. *J. Cogn. Neurosci.* **1993**, *5*, 129–149. [CrossRef]
105. Tommasi, L.; Bressan, P.; Vallortigara, G. Solving Occlusion Indeterminacy in Chromatically Homogeneous Patterns. *Perception* **1995**, *24*, 391–403. [CrossRef]

106. Pinna, B.; Reeves, A.; Koenderink, J.; van Doorn, A.; Deiana, K. A new principle of figure-ground segregation: The accentuation. *Vis. Res.* **2018**, *143*, 9–25. [CrossRef]
107. Pinna, B.; Sirigu, L. The Accentuation Principle of Visual Organization and the Illusion of Musical Suspension. *See. Perce.* **2011**, *24*, 595–621. [CrossRef]
108. Pinna, B.; Sirigu, L. New visual phenomena and the illusion of musical downbeat: Space and time organization due to the accentuation principle. *Acta Psychologica.* **2016**, *170*, 32–57. [CrossRef]
109. Pinna, B.; Reeves, A. On the purposes of color for living beings: toward a theory of color organization. *Psychol. Res.* **2015**, *79*, 64–82. [CrossRef] [PubMed]
110. Pinna, B.; Porcheddu, D.; Deiana, K. From Grouping to Coupling: A New Perceptual Organization in Vision, Psychology, and Biology. *Front. Psychol.* **2016**, *7*, 1119. [CrossRef] [PubMed]
111. D'Angiulli, A.; Reeves, A. The relationship between self-reported vividness and latency during mental size scaling of everyday items: Phenomenological evidence of different types of imagery. *Am. J. Psychol.* **2007**, *120*, 51–521. [CrossRef] [PubMed]
112. Alter, A.L.; Balcetis, E. Fondness makes the distance grow shorter: desired locations seem closer because they seem more vivid. *J. Exp. Soc. Psychol.* **2010**, *47*, 16–21. [CrossRef]
113. Barrowcliff, A.L.; Gray, N.S.; Freeman, T.C.; MacCulloch, M.J. Eye-movements reduce the vividness, emotional valence and electrodermal arousal associated with negative autobiographical memories. *J. Forensic Psychiatry Psychol.* **2004**, *15*, 325–345. [CrossRef]
114. Bywaters, M.; Andrade, J.; Turpin, G. Determinants of the vividness of visual imagery: The effects of delayed recall, stimulus affect and individual differences. *Memory* **2004**, *12*, 479–488. [CrossRef]
115. Baddeley, A.D.; Andrade, J. Working memory and the vividness of imagery. *J. Exp. Psychol. Gen.* **2000**, *129*, 126–145. [CrossRef]
116. Cattaneo, Z.; Bona, S.; Silvanto, J. Cross-adaptation combined with TMS reveals a functional overlap between vision and imagery in the early visual cortex. *NeuroImage* **2012**, *59*, 3015–3020. [CrossRef]
117. Cattaneo, Z.; Pisoni, A.; Papagno, C.; Silvanto, J. Modulation of Visual Cortical Excitability by Working Memory: Effect of Luminance Contrast of Mental Imagery. *Front. Psychol.* **2011**, *2*, 29. [CrossRef]
118. Cui, X.; Jeter, C.B.; Yang, D.; Montague, P.R.; Eagleman, D.M. Vividness of mental imagery: Individual variability can be measured objectively. *Vis. Res.* **2007**, *47*, 474–478. [CrossRef]
119. Farah, M.J.; Peronnet, F. Event-related potentials in the study of mental imagery. *J. Psychophysiol.* **1989**, *3*, 99–109.
120. Farah, M.J.; Weisberg, L.L.; Monheit, M.; Peronnet, F. Brain Activity Underlying Mental Imagery: Event-related Potentials During Mental Image Generation. *J. Cogn. Neurosci.* **1989**, *1*, 302–316. [CrossRef] [PubMed]
121. Sparing, R.; Mottaghy, F.M.; Ganis, G.; Thompson, W.L.; Töpper, R.; Kosslyn, S.M.; Pascual-Leone, A. Visual cortex excitability increases during visual mental imagery—A TMS study in healthy human subjects. *Brain Res.* **2002**, *938*, 92–97. [CrossRef]
122. D'Angiulli, A. Mental image generation and the contrast sensitivity function. *Cognition* **2002**, *85*, B11–B19. [CrossRef]
123. D'Angiulli, A. Is the spotlight an obsolete metaphor of "seeing with the mind's eye"? A constructive naturalistic approach to the inspection of visual mental images. *Imagin. Cogn. Pers.* **2008**, *28*, 35–117. [CrossRef]
124. D'Angiulli, A.; Reeves, A. Probing vividness increment through imagery interruption. *Imagin. Cogn. Pers.* **2004**, *23*, 79–88. [CrossRef]

Comment

Dubious Claims about Simplicity and Likelihood: Comment on Pinna and Conti (2019)

Peter A. van der Helm

Department of Brain & Cognition, University of Leuven (K.U. Leuven), Tiensestraat 102-box 3711, B-3000 Leuven, Belgium; peter.vanderhelm@kuleuven.be

Received: 17 September 2019; Accepted: 13 January 2020; Published: 16 January 2020

Abstract: Pinna and Conti (*Brain Sci.*, 2019, *9*, 149, doi:10.3390/brainsci9060149) presented phenomena concerning the salience and role of contrast polarity in human visual perception, particularly in amodal completion. These phenomena are indeed illustrative thereof, but here, the focus is on their claims (1) that neither simplicity nor likelihood approaches can account for these phenomena; and (2) that simplicity and likelihood are equivalent. I argue that their first claim is based on incorrect assumptions, whereas their second claim is simply untrue.

Keywords: contrast polarity; perceptual organization; simplicity principle; likelihood principle; simplicity–likelihood equivalence; Bayes; classical information theory; modern information theory

---

1. Introduction

In the context of this journal's special issue *Vividness, consciousness, and mental imagery: Making the missing links across disciplines and methods* [1], Pinna and Conti [2] presented fine phenomena concerning the salience and role of contrast polarity in human visual perception, particularly in amodal completion. They also claimed (1) that these phenomena go against existing simplicity and likelihood approaches to visual perception, and (2) that simplicity and likelihood are equivalent. Before they submitted their article to this journal, however, they had been informed that these claims are incorrect—a matter, by the way, of formal facts rather than psychological opinions. To set the stage, I first sketch the perceptual topic of their study.

2. Contrast Polarity

The role of contrast polarity in human visual perception is a long-standing research topic (see, e.g., in [3–7]). As a consequence, the phenomena presented by Pinna and Conti are not really as novel or surprising as they suggested them to be, but they are indeed illustrative of the effects of, in particular, contrast polarity reversals. For example, Figure 1 shows, in the style of Pinna and Conti, a stimulus in which such a reversal triggers a substantial change in the way in which it is perceptually organized.

Depending on stimulus type, contrast polarity also affects visual regularity detection. For instance, the evident symmetry in the checkerboard pattern in Figure 2a is perceptually destroyed by the contrast polarity reversal in its right-hand half in Figure 2b (see, e.g., in [9]). In dot patterns, however, such a reversal does not seem to do much harm (see, e.g., in [10–12]). Furthermore, Figure 2c depicts a rotational Glass pattern with dipoles consisting of either two black dots or two white dots. It exhibits a moiré effect that is perceptually about as strong as when all dipoles consist of black dots (personal observation). However, when all dipoles are identical with one black dot and one white dot, as depicted in Figure 2d, the moiré effect disappears (see, e.g., [4,13,14]). The moiré effect does not disappear, by the way, when every dipole consists of differently shaped elements [15].

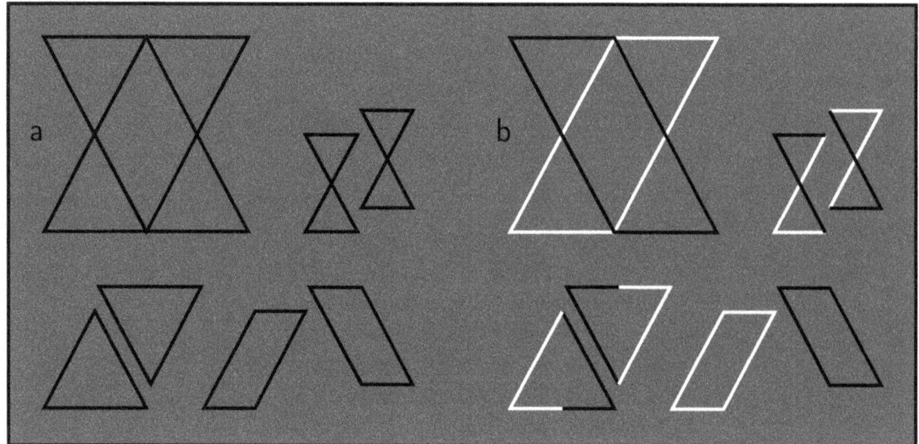

Figure 1. Contrast polarity and perceptual organization. Both compound stimuli can be described (i.e., interpreted) as consisting of, for instance, two triangles, two diabolos, or two parallelograms. The parallelogram interpretation is probably the weakest one in panel (**a**); but due to contrast polarity reversals, it is definitely the strongest one in panel (**b**) (after [8]).

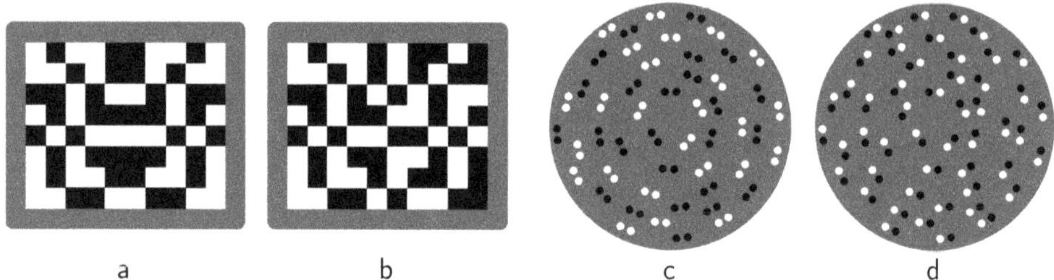

Figure 2. Contrast polarity and regularity detection. The symmetry in the checkerboard pattern in panel (**a**) is perceptually destroyed by the contrast polarity reversal in its right-hand half in panel (**b**); The Glass pattern in panel (**c**), with dipoles consisting of either two black or two white dots, exhibits a clear moiré effect, which disappears in panel (**d**), where all dipoles are identical with one black dot and one white dot.

Therefore, I agree with Pinna and Conti that contrast polarity is a factor to reckon with. However, I think that, in their first claim, they concluded too easily—in fact, based on incorrect assumptions—that contrast polarity triggers local groupings that precede the global groupings allegedly predicted by simplicity and likelihood approaches. Next, this is discussed in more detail.

3. Incorrect Assumptions

The following three quotes from Pinna and Conti illustrate their first claim, that is, their stance about contrast polarity versus simplicity and likelihood approaches:

> "The salience and visibility, derived by the largest amplitude of luminance dissimilarity imparted by contrast polarity, precedes any holist or likelihood organization due to simplicity/Prägnanz and Bayes' inference." [2] (p. 12 of 32)

> "Contrast polarity was shown to operate locally, eliciting results that could be independent from any global scale and that could also be paradoxical. These results weaken and challenge theoretical approaches based on notions like oneness, unitariness, symmetry, regularity, simplicity, likelihood, priors, constraints, and past knowledge. Therefore, Helmholtz's likelihood principle,

> simplicity/Prägnanz, and Bayes' inference were clearly questioned since they are supposed to operate especially at a global and holistic level of vision." [2] (p. 26 of 32)
>
> "The highlighting strength of contrast polarity determines even the grouping effectiveness against the global and holistic rules and factors expected by Helmholtz's likelihood principle, simplicity/Prägnanz, and Bayes' inference." [2] (p. 26 of 32)

It is true that simplicity and likelihood approaches may aim to arrive at global stimulus interpretations, but a general objection against the above stance is that they (can) do so by including local factors as well. For instance, van Lier [16] presented a theoretically sound and empirically adequate simplicity model for the integration of global and local aspects in amodal completion (see also [17]). A methodological objection is that Pinna and Conti introduced contrast polarity changes in stimuli but pitted these against alleged simplicity and likelihood predictions for the unchanged stimuli. As I specify next, this is unfair, and in my view, scientifically inappropriate.

3.1. Likelihood

Probabilistic approaches in cognitive science, and beyond, span a large spectrum (see Table 1, upper part, for a first impression thereof). At one end of this spectrum are Bayesian approaches like Friston's free-energy predictive coding approach [18]. This approach claims to have high explanatory power, but in fact, hardly goes beyond data accommodation, hardly produces falsifiable predictions, and suffers from computational intractability (see, e.g., [19–23]). In my view, it therefore qualifies as what Chomsky called an "analogic guess", that is, it "creates the illusion of a rigorous scientific theory with very broad scope" [24] (p. 32).

Most Bayesian models in cognitive science, however, take a stance that, in my view, is more appropriate and adequate. Instead of considering the likelihood principle as a strong explanatory principle, they rather consider it as a powerful modeling principle by which free-to-choose probabilities can be assigned to free-to-choose things in order to perform sophisticated data fitting and data extrapolation. This may not always be easy to do, but it means that there is no fundamental obstacle for the inclusion of local aspects like the effects of contrast polarity. Pinna and Conti wrote "[If] we do not consider the contrast polarity as a constraint or as a prior, Bayes' inference cannot easily explain these conditions" [2] (p. 12 of 32)—indeed, but why would we? Therefore, they knowingly ignored the above flexibility and applied likelihood as if it is fundamentally blind to contrast polarity. Thereby, they missed the mark in their assessment of likelihood approaches.

3.2. Simplicity

Compared to the likelihood principle, the simplicity principle is less of a modeling principle and more of an explanatory principle. By this, I do not mean to claim that simplicity explains all contrast polarity phenomena. For instance, in Glass patterns, simplicity predicts stronger moiré effects for identical dipoles than for nonidentical ones, which may often be adequate but, as indicated, not in the case of Figure 2c,d. The point is that I consider simplicity to be a fundamental force in perception, which nevertheless—just as gravity in physics, for instance—interacts with other forces, yielding results that now and again may deviate from what simplicity on its own would yield.

In this respect, notice that the contrast polarity reversal in Figure 2b can be said to trigger local groupings which destroy the symmetry. It can also be said, however, to yield antisymmetry, which, on formal theoretical grounds within the simplicity paradigm, is predicted to be not one of the instantaneously detectable visual regularities [25,26]. The earlier mentioned reversal in dot patterns seems an exception to both rules [9]. Furthermore, the reversal in Figure 1 clearly implies that the parallelogram interpretation becomes less complex compared to the other two interpretations—provided one applies, unlike Pinna and Conti did, the simplicity idea correctly. This idea is specified next in some more detail (see Table 1, lower part, for a first impression thereof).

In both mathematics and perception research, the simplicity idea falls within a descriptive framework. It relies on regularity extraction to obtain simplest descriptive codes, which capture hierarchical organizations of given input. There are a few domain-dependent differences (see Table 1). Unlike algorithmic information theory in mathematics does, structural information theory in perception research employs a fixed descriptive coding language extracting theoretically and empirically grounded visual regularities, and it classifies things on the basis of the hierarchical organizations described by simplest descriptive codes (which are taken to reflect mental representations) [27–35].

Table 1. Overview of probabilistic and descriptive frameworks.

Probabilistic framework
Classical information theory:
• information load of thing with probability p is surprisal $-\log(p)$ [36,37] (information quantified by its probability, not by its content) • codes are nominalistic labels referring to things (as, e.g., in the Morse Code) • optimal coding by label codes the length of surprisals [38] (implying minimal long-term average code length, not individually shortest codes)
Likelihood principle [39]:
• preference for things with higher probabilities, i.e., with lower surprisals
Bayesian inference (*incomputable* [21]):
• in cognitive science: free-to-choose probabilities for free-to-choose things • minimum message length principle (message length measured i.t.o. surprisals) [40]
Surprisals do not enable descriptive formulation
Descriptive framework
Modern information theory (triggered by the question: what if probabilities are unknown?):
• codes are hierarchical descriptions (i.e., reconstruction recipes) of individual things • shorter descriptive codes by extracting regularities i.t.o. identity relationships between parts • information load of thing is its complexity C, i.e., the length of its shortest descriptive code (information quantified by its content, not by its probability)
Simplicity principle:
• a.k.a. minimum principle [41] or minimum description length principle [42] • preference for simpler things, i.e., things with shorter descriptive codes
Algorithmic information theory (*mathematics*) [43–45]:
• extraction of any imaginable regularity (*incomputable*) • classification by complexity of simplest descriptive code
Structural information theory (*cognitive science*) [27–29]:
• extraction of theoretically and empirically grounded visual regularities (*computable*) [30–33] • classification by hierarchical organization described by simplest descriptive code [34,35]
Precisals 2^{-C}, a.k.a. algorithmic probabilities, enable probabilistic formulation

A shared point, however, is that descriptive codes constitute reconstruction recipes for stimuli (just as computer algorithms are reconstruction recipes for the output they produce). Therefore, if a stimulus contains different contrast polarities, then these are necessarily also accounted for by descriptive codes of this stimulus. In Figure 1b, for instance, this implies that the contrast polarity changes in the triangles and diabolos make them more complex than the parallelograms without such changes are. Pinna and Conti knowingly ignored this and applied simplicity as if it is fundamentally blind to contrast polarity. Thereby, they missed the mark in their assessment of simplicity approaches.

3.3. Summary (1)

My objective here was not to show that simplicity and likelihood approaches can account for all contrast polarity phenomena (on their own, they certainly can account for some but probably not for all). Instead, my objective was to show that Pinna and Conti applied these approaches incorrectly, even though they had been warned about this. Thereby, they knowingly ignored that these approaches are far more flexible than they assumed them to be. In my view, this is scientifically inappropriate.

4. Simplicity and Likelihood Are Not Equivalent

Pinna and Conti formulated their second claim, about the alleged equivalence of simplicity and likelihood, as follows.

> "[...] the visual object that minimizes the description length is the same one that maximizes the likelihood. In other terms, the most likely hypothesis about the perceptual organization is also the outcome with the shortest description of the stimulus pattern." [2] (p. 3 of 32)

This is an extraordinary claim. It therefore requires extraordinary evidence, but Pinna and Conti actually provided no corroboration at all (in their earlier draft, they cited Chater [46]; see Section 4.1. Instead, they seem to have jumped on the bandwagon of an idea that, for the past 25 years, has lingered on in the literature—in spite of refutations. As said, Pinna and Conti had been informed about its falsehood but chose to persist. It is therefore expedient to revisit the alleged equivalence of simplicity and likelihood (see Table 1 for a synopsis of relevant issues and terminologies).

Before going into specific equivalence claims, I must say that, to me, it is hard to even imagine that simplicity and likelihood might be equivalent. Notice that descriptive simplicity is a fairly stable concept. That is, as has been proved in modern information theory (IT) in mathematics, every reasonable descriptive coding language yields about the same complexity ranking for things [43–45]. Probabilities, conversely, come in many shapes and forms. For instance, on the one hand, in technical contexts like communication theory, the to-be-employed probabilities may be (approximately) known—though notice that they may vary with the situation at hand. For known probabilities, one may aim at minimal long-term average code length for large sets of identical and nonidentical messages (i.e., Shannon's [38] optimal coding), and by the same token, at compounds of label codes that yield data compression for large compounds of identical and nonidentical messages (see, e.g., in [47,48]). On the other hand, the Helmholtzian likelihood principle in perception is now and again taken to rely on objective "real" probabilities of things in the world. This would give it an explanatory nature, but by all accounts, it seems impossible to assess such probabilities (see, e.g., in [49,50]). In between are, for instance, Bayesian models in cognitive science. In general, as said, such models employ free-to-choose probabilities for free-to-choose things, where both those things and their probabilities may be chosen subjectively on the basis of experimental data or modeller's intuition. Therefore, all in all, how could one ever claim that fairly stable descriptive complexities are equivalent to every set of probabilities employed or proposed within the probabilistic framework?

Yet, notice that Pinna and Conti are not alone in their equivalence claim. Equivalence also has been claimed by, for instance, Friston [51], Feldman [52,53], and Thornton [54]. They too failed to provide explicit corroboration, which raises the question of where the claim actually comes from. As a matter of fact, for alleged support, they all referred consistently to either Chater [46] or MacKay [55], or to both. These sources are discussed next (for more details, see in [17,22,56,57]).

4.1. Chater (1996)

The main issue in the well-cited article by Chater [46] may be explained by means of Figure 3, starting at the left-hand side. The upper-left quadrant indicates that, for some set of probabilities p, one can maximize certainty via Bayes' rule, that is, by combining prior probabilities $p(H)$ and conditional probabilities $p(D|H)$ for data D and hypotheses H to obtain posterior probabilities $p(H|D)$. [*Note:* in general, priors account for viewpoint-independent aspects (i.e., how good is hypothesis H in itself?), whereas conditionals account for viewpoint-dependent aspects (i.e., how well do data D fit hypothesis H?).] The lower-left quadrant indicates information measurement in the style of classical IT, that is, by the conversion of probabilities p to surprisals $I = -\log p$ (term coined by Tribus [58]; concept developed by Nyquist [36] and Hartley [37]). As said, the surprisal can be used to achieve optimal coding [38], but as indicated in Figure 3, prior and conditional surprisals can, analogous to Bayes' rule, also be combined to minimize information as quantified in classical IT. The latter constitutes the minimal message length principle (MML) [40], which, considering the foregoing, clearly is a full Bayesian approach that merely has been rewritten in terms of surprisals [59].

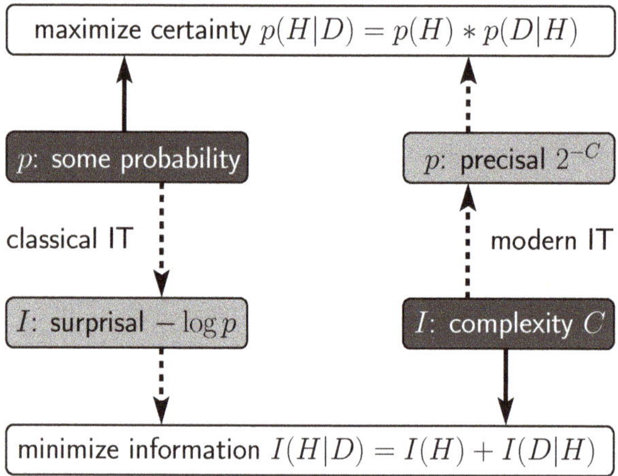

Figure 3. Surprisals versus precisals. For data D and hypotheses H, probabilities p can be used to maximize Bayesian certainty under these probabilities (top left), and via the surprisal conversion $I = -\log p$, also to minimize information as quantified in classical information theory (IT) (bottom left). Descriptive complexities C can be used to minimize information as quantified in modern IT (bottom right), and via the precisal conversion $p = 2^{-C}$, also to maximize Bayesian certainty under these precisals (top right) (adapted from [57]).

Turning to the right-hand side of Figure 3, the lower-right quadrant indicates that, for some descriptive coding language yielding complexities C, one can combine prior and conditional complexities to minimize information as quantified in modern IT. This is the minimum description length principle (MDL) [42], which can be seen as a modern version of Occam's razor [60]. It also reflects the current take on the simplicity principle in perception [16,17]. The upper-right quadrant indicates that complexities C can be converted to what are called algorithmic probabilities $p = 2^{-C}$, also called precisals [17]. These are artificial probabilities but, just as holds for other probabilities, prior and conditional precisals can, for instance, be combined to maximize certainty via Bayes' rule. This reflects Solomonoff's [44,45] Leitmotif: because classical IT relies on known probabilities, he wondered if one could devise "universal" probabilities, that is, probabilities that can be used fairly reliably whenever the actual probabilities are unavailable. In modern IT, precisals are proposed to be such universal probabilities and much research goes into their potential reliability. In cognitive science, they can be used, for instance, to predict the likelihood of empirical outcomes according to simplicity (i.e., rather than assuming that the brain itself uses them to arrive at those outcomes).

The surprisal and precisal conversions are convenient in that they allow for sophisticated theoretical comparisons between simplicity and likelihood approaches (see, e.g., in [59,60]). Chater, however, jumped to the conclusion that these conversions imply that simplicity and likelihood are equivalent. Notice that the left-hand and right-hand sides in Figure 3 represent fundamentally different starting points and lines of reasoning. Therefore, equivalence would hold only if, in the lower half, the left-hand probability-based quantification of information and the right-hand content-based quantification of information—or, in the upper half, the related left-hand and right-hand sets of probabilities—are identical. Apart from the fundamental questionability thereof, these were not issues Chater addressed. It is true that the conversions imply that simplicity and likelihood can use the same minimization and maximization formulas, but Chater fatally overlooked that equivalence depends crucially on what they substitute in those formulas—here, it is clear that they substitute fundamentally different things. Chater's mistake is in fact like claiming that Newton's formula ma for force F is equivalent to Einstein's formula mc^2 for energy E—allegedly because both could have used a formula like mX, but fatally ignoring that X is something fundamentally different in each case. Therefore, all in all, Chater provided no evidence for equivalence of simplicity and likelihood at all.

4.2. MacKay (2003)

In what soon became a standard Bayesian textbook, MacKay [55] devoted one chapter (Chapter 28) to links between simplicity and likelihood. He actually did not claim equivalence, but as I discussed in [57] and revisit here, he mistakenly equated surprisals and description lengths, and he made an admittedly compelling argument that, however, was overinterpreted by others—who, subsequently, did claim equivalence.

One of MacKay's conclusions was that "MDL has no apparent advantages over the direct probabilistic approach" [55] (p. 352). However, he attributed MDL not to MDL developer Rissanen [42] but to MML developers Wallace and Boulton [40]—just as [61] later did too, by the way. In fact, in the entire chapter, Mackay mistakenly wrote "MDL" instead of "MML" and "description length" instead of "message length" or "surprisal" (Baxter & Oliver [62] noticed this mistake also in MacKay [63]). Therefore, he in fact discussed the Bayesian MML and not the non-Bayesian MDL. No wonder, therefore, that he saw "no apparent advantages". Unfortunately, his mistake added to the already existing misconceptions surrounding simplicity and likelihood. For instance, subsequently, Feldman [53,64–67] also mixed up MDL's description lengths (which, i.t.o. modern IT's descriptive codes, aim at minimal code length for individual things) and MML's surprisals (which, i.t.o. classical IT's label codes, minimize long-term average code length for large sets of identical, and nonidentical things).

MacKay's mistake above already may have triggered equivalence claims, but unintentionally, another conclusion may have done so more strongly. That is, he also argued that "coherent inference (as embodied by Bayesian probability) automatically embodies Occam's razor" [55] (p. 344). This is easily read as suggesting equivalence (see, e.g., in [52,53]), but notice that MacKay reasoned as follows.

> "Simple models tend to make precise predictions. Complex models, by their nature, are capable of making a greater variety of predictions [...]. So if H_2 is a more complex model [than H_1], it must spread its predictive probability $P(D|H_2)$ more thinly over the data space than H_1. Thus, in the case where the data are compatible with both theories, the simpler H_1 will turn out more probable than H_2, without our having to express any subjective dislike for complex models." [55] (p. 344)

In other words, he argued that conditional probabilities, as used in Bayesian modeling, show a bias towards hypotheses with low prior complexity. This is definitely interesting and compelling, and as he noted, it reveals subtle intricacies in Bayesian inference.

Currently relevant, however, is that it does not imply equivalence of simplicity and likelihood. For instance, regarding both priors and conditionals, it is silent about how close (fairly stable) simplicity-based precisals and (fairly flexible) Bayesian probabilities might be. Furthermore, whereas prior precisals are nonuniform by nature, MacKay explicitly assumed uniform prior probabilities (he needs this not-truly-Bayesian assumption, because nonuniform prior probabilities could easily overrule the bias he attributed to conditional probabilities). This assumption as such already excludes equivalence. Notice further that he neither gave a formal definition of complexity nor a formal proof of his argument. This means that his argument, though certainly compelling, does not reflect a formally proven fact. Thereby, it has the same status as, for instance, van der Helm's [17] argument that, specifically in visual perceptual organization, simplicity-based conditional precisals are close to intuitively real conditional probabilities—which would imply that precisals are fairly reliable in the everyday perception by moving observers. It is true that both arguments reflect interesting rapprochements between simplicity and likelihood, but neither argument asserts equivalence.

4.3. Summary (2)

My objective here was to trace back where Pinna and Conti's misguided equivalence claim came from. This led to Chater [46] and MacKay [55], whose flawed comprehension of the links between classical IT and modern IT seems to have given rise to various misconceptions. It is true that they pointed at interesting things, but they did not provide any evidence of equivalence of simplicity and likelihood. With fundamentally different baits, classical IT and modern IT are fishing in the same pond of probabilities and information measurements—using a perhaps mind-boggling body of terms. It is therefore understandable that comparisons between them may be confusing, particularly to those who are less trained in formal reasonings. Persisting in an equivalence claim after having been informed in detail that such a claim is nonsense—as Pinna and Conti did—is another matter however, and in my view, scientifically inappropriate.

5. Conclusions

In this comment, I revisited the claims put forward by Pinna and Conti. First, they argued that simplicity and likelihood approaches cannot account for the contrast polarity phenomena they presented. I showed, however, that their argument was based on incorrect assumptions and that simplicity and likelihood approaches are far more flexible than they assumed them to be—without claiming, by the way, that they can account for all contrast polarity phenomena. Second, even though it did not seem essential in their article, they argued that simplicity and likelihood are equivalent. I showed, however, that, although this issue is prone to confusion, there is no reason whatsoever to suppose that simplicity and likelihood might be equivalent. Considering that this is a matter of formal facts rather than psychological opinions, it is, in my view, worrying that—in spite of refutations—unsubstantiated equivalence claims linger on in the literature.

Funding: This research was supported by Methusalem grant METH/14/02 awarded to Johan Wagemans (www.gestaltrevision.be).

Conflicts of Interest: The author declares no conflict of interest.

References

1. D'Angiulli, A. (Ed.) *Special Issue "Vividness, Consciousness, And Mental Imagery: Making The Missing Links Across Disciplines and Methods"*; Brain Science; MDPI: Basel, Switzerland, 2019; Volume 9.
2. Pinna, B.; Conti, L. The limiting case of amodal completion: The phenomenal salience and the role of contrast polarity. *Brain Sci.* **2019**, *9*, 149. [CrossRef]
3. Bell, J.; Gheorghiu, E.; Hess, R.F.; Kingdom, F.A.A. Global shape processing involves a hierarchy of integration stages. *Vision Res.* **2011**, *51*, 1760–1766. [CrossRef] [PubMed]
4. Elder, J.; Zucker, S. The effect of contour closure on the rapid discrimination of two-dimensional shapes. *Vision Res.* **1993**, *33*, 981–991. [CrossRef]
5. Schira, M.M.; Spehar, B. Differential effect of contrast polarity reversals in closed squares and open L-junctions. *Front. Psychol. Perception Sci.* **2011**, *2*, 47. [CrossRef] [PubMed]
6. Spehar, B. The role of contrast polarity in perceptual closure. *Vision Res.* **2002**, *42*, 343–350. [CrossRef]
7. Su, Y.; He, Z.J.; Ooi, T.L. Surface completion affected by luminance contrast polarity and common motion. *J. Vision* **2010**, *10*, 1–14. [CrossRef] [PubMed]
8. Reed, S.K. Structural descriptions and the limitations of visual images. *Mem. Cogn.* **1974**, *2*, 329–336. [CrossRef] [PubMed]
9. Mancini, S.; Sally, S.L.; Gurnsey, R. Detection of symmetry and antisymmetry. *Vision Res.* **2005**, *45*, 2145–2160. [CrossRef]
10. Saarinen, J.; Levi, D.M. Perception of mirror symmetry reveals long-range interactions between orientation-selective cortical filters. *Neuroreport* **2000**, *11*, 2133–2138. [CrossRef]

11. Tyler, C.W.; Hardage, L. Mirror symmetry detection: Predominance of second-order pattern processing throughout the visual field. In *Human Symmetry Perception and Its Computational Analysis*; Tyler, C.W., Ed.; VSP: Zeist, The Netherlands, 1996; pp. 157–172.
12. Wenderoth, P. The effects of the contrast polarity of dot-pair partners on the detection of bilateral symmetry. *Perception* **1996**, *25*, 757–771. [CrossRef]
13. Wilson, J.A.; Switkes, E.; De Valois, R.L. Glass pattern studies of local and global processing of contrast variations. *Vision Res.* **2004**, *44*, 2629–2641. [CrossRef] [PubMed]
14. Or, C.C.-F.; Khuu, S.K.; Hayes, A. The role of luminance contrast in the detection of global structure in static and dynamic, same- and opposite-polarity, Glass patterns. *Vision Res.* **2007**, *47*, 253–259. [CrossRef] [PubMed]
15. Prazdny, K. On the perception of Glass patterns. *Perception* **1984**, *13*, 469–478. [CrossRef]
16. van Lier, R.J.; van der Helm, P.A.; Leeuwenberg, E.L.J. Integrating global and local aspects of visual occlusion. *Perception* **1994**, *23*, 883–903. [CrossRef] [PubMed]
17. van der Helm, P.A. Simplicity versus likelihood in visual perception: From surprisals to precisals. *Psychol. Bull.* **2000**, *126*, 770–800. [CrossRef] [PubMed]
18. Friston, K. The free-energy principle: A rough guide to the brain? *Trends Cogn. Sci.* **2009**, *13*, 293–301. [CrossRef] [PubMed]
19. Luccio, R. Limits of the application of Bayesian modeling to perception. *Perception* **2019**, *48*, 901–917. [CrossRef] [PubMed]
20. Radomski, B.M. The Theoretical Status of the Free-Energy Principle. Ph.D. Thesis, Ruhr-University Bochum, Bochum, Germany, 2019.
21. Kwisthout, J.; van Rooij, I. Computational resource demands of a predictive Bayesian brain. *Comput. Brain Behav.* **2019**. [CrossRef]
22. van der Helm, P.A. Structural coding versus free-energy predictive coding. *Psychon. B. Rev.* **2016**, *23*, 663–677. [CrossRef]
23. Wang, P. The limitation of Bayesianism. *Artif. Intell. 158* **2004**, *1*, 97–106. [CrossRef]
24. Chomsky, N. A review of B. F. Skinner's *Verbal Behavior*. *Language 35* **1959**, *1*, 26–58. [CrossRef]
25. van der Helm, P.A. *Simplicity in Vision: A Multidisciplinary Account of Perceptual Organization*; Cambridge University Press: Cambridge, UK, 2014.
26. van der Helm, P.A.; Treder, M.S. Detection of (anti)symmetry and (anti)repetition: Perceptual mechanisms versus cognitive strategies. *Vision Res.* **2009**, *49*, 2754–2763. [CrossRef]
27. Leeuwenberg, E.L.J. *Structural Information of Visual Patterns: An Efficient Coding System in Perception*; Mouton & Co.: Hague, The Netherlands, 1968.
28. Leeuwenberg, E.L.J. Quantitative specification of information in sequential patterns. *Psychol. Rev.* **1969**, *76*, 216–220. [CrossRef]
29. Leeuwenberg, E.L.J. A perceptual coding language for visual and auditory patterns. *Am. J. Psychol.* **1971**, *84*, 307–349. [CrossRef] [PubMed]
30. van der Helm, P.A.; Leeuwenberg, E.L.J. Accessibility, a criterion for regularity and hierarchy in visual pattern codes. *J. Math. Psychol.* **1991**, *35*, 151–213. [CrossRef]
31. Van der Helm, P.A.; Leeuwenberg, E.L.J. Goodness of visual regularities: A nontransformational approach. *Psychol. Rev.* **1996**, *103*, 429–456. [CrossRef] [PubMed]
32. Makin, A.D.J.; Wright, D.; Rampone, G.; Palumbo, L.; Guest, M.; Sheehan, R.; Cleaver, H.; Bertamini, M. An electrophysiological index of perceptual goodness. *Cereb. Cortex* **2016**, *26*, 4416–4434. [CrossRef]
33. van der Helm, P.A. Transparallel processing by hyperstrings. *Proc. Natl. Acad. Sci. USA* **2004**, *101*, 10862–10867. [CrossRef]
34. Leeuwenberg, E.L.J.; van der Helm, P.A. Unity and variety in visual form. *Perception* **1991**, *20*, 595–622. [CrossRef]
35. Leeuwenberg, E.L.J.; van der Helm, P.A.; van Lier, R.J. From geons to structure: A note on object classification. *Perception* **1994**, *23*, 505–515. [CrossRef]
36. Nyquist, H. Certain factors affecting telegraph speed. *Bell Syst. Tech. J.* **1924**, *3*, 324–346. [CrossRef]
37. Hartley, R.V.L. Transmission of information. *Bell Syst. Tech. J.* **1928**, *7*, 535–563. [CrossRef]
38. Shannon, C.E. A mathematical theory of communication. *Bell Syst. Tech. J.* **1948**, *27*, 623–656. [CrossRef]

39. von Helmholtz, H.L.F. *Treatise on Physiological Optics*; Translated by Southall, J.P.C.; Original work published 1909; Dover: New York, NY, USA, 1962.
40. Wallace, C.; Boulton, D. An information measure for classification. *Comput. J.* **1968**, *11*, 185–194. [CrossRef]
41. Hochberg, J.E.; McAlister, E. A quantitative approach to figural "goodness". *J. Exp. Psychol.* **1953**, *46*, 361–364. [CrossRef] [PubMed]
42. Rissanen, J. Modelling by the shortest data description. *Automatica* **1978**, *14*, 465–471. [CrossRef]
43. Kolmogorov, A.N. Three approaches to the quantitative definition of information. *Probl. Inform. Transm.* **1965**, *1*, 1–7. [CrossRef]
44. Solomonoff, R.J. A formal theory of inductive inference, Part 1. *Inform. Control* **1964**, *7*, 1–22. [CrossRef]
45. Solomonoff, R.J. A formal theory of inductive inference, Part 2. *Inform. Control* **1964**, *7*, 224–254. [CrossRef]
46. Chater, N. Reconciling simplicity and likelihood principles in perceptual organization. *Psychol. Rev.* **1996**, *103*, 566–581. [CrossRef]
47. Hyvärinen, A.; Hurri, J.; Hoyer, P.O. *Natural Image Statistics*; Springer: London, UK, 2009.
48. Renoult, J.P.; Mendelson, T.C. Processing bias: Extending sensory drive to include efficacy and efficiency in information processing. *Proc. R. Soc. B* **2019**, *286*, 20190445. [CrossRef]
49. Feldman, J. Tuning your priors to the world. *Top. Cogn. Sci.* **2013**, *5*, 13–34. [CrossRef]
50. Hoffman, D.D. What do we mean by "The structure of the world"? In *Perception as Bayesian Inference*; Knill, D.C., Richards, W., Eds.; Cambridge University Press: Cambridge, MA, USA, 1996; pp. 219–221.
51. Friston, K.; Chu, C.; Mourão-Miranda, J.; Hulme, O.; Rees, G.; Penny, W.; Ashburner, J. Bayesian decoding of brain images. *NeuroImage* **2008**, *39*, 181–205. [CrossRef] [PubMed]
52. Feldman, J. Bayes and the simplicity principle in perception. *Psychol. Rev.* **2009**, *116*, 875–887. [CrossRef] [PubMed]
53. Feldman, J. The simplicity principle in perception and cognition. *WIREs Cogn. Sci.* **2016**, *7*, 330–340. [CrossRef] [PubMed]
54. Thornton, C. Infotropism as the underlying principle of perceptual organization. *J. Math. Psychol.* **2014**, *61*, 38–44. [CrossRef]
55. MacKay, D.J.C. *Information Theory, Inference, and Learning Algorithms*; Cambridge University Press: Cambridge, UK, 2003.
56. van der Helm, P.A. Bayesian confusions surrounding simplicity and likelihood in perceptual organization. *Acta Psychol.* **2011**, *138*, 337–346. [CrossRef]
57. van der Helm, P.A. On Bayesian simplicity in human visual perceptual organization. *Perception* **2017**, *46*, 1269–1282. [CrossRef]
58. Tribus, M. *Thermostatics and Thermodynamics*; Van Nostrand: Princeton, NJ, USA, 1961.
59. Grünwald, P.D. *The Minimum Description Length Principle*; MIT Press: Cambridge, MA, USA, 2007.
60. Li, M.; Vitányi, P. *An Introduction to Kolmogorov Complexity and Its Applications*, 2nd ed.; Springer: New York, NY, USA, 1997.
61. Penny, W.D.; Stephan, K.E.; Mechelli, A.; Friston, K.J. Comparing dynamic causal models. *NeuroImage* **2004**, *22*, 1157–1172. [CrossRef]
62. Baxter, R.A.; Oliver, J.J. *MDL and MML: Similarities and Differences*; Tech Report 207; Monash University: Melbourne, Australia, 1994.
63. MacKay, D.J.C. Bayesian Methods for Adaptive Models. Ph.D. Thesis, California Institute of Technology, Pasadena, CA, USA, 1992.
64. Feldman, J.; Singh, M. Bayesian estimation of the shape skeleton. *Proc. Natl. Acad. Sci. USA* **2006**, *103*, 18014–18019. [CrossRef]
65. Froyen, V.; Feldman, J.; Singh, M. Bayesian hierarchical grouping: Perceptual grouping as mixture estimation. *Psychol. Rev.* **2015**, *122*, 575–597. [CrossRef] [PubMed]

66. Wilder, J.; Feldman, J.; Singh, M. Contour complexity and contour detection. *J. Vision* **2015**, *15*, 1–16. [CrossRef] [PubMed]
67. Wilder, J.; Feldman, J.; Singh, M. The role of shape complexity in the detection of closed contours. *Vision Res.* **2016**, *126*, 220–231. [CrossRef] [PubMed]

Reply

On the Role of Contrast Polarity: In Response to van der Helm's Comments

Baingio Pinna [1,*] **and Livio Conti** [2]

[1] Department of Biomedical Science, University of Sassari, 07100 Sassari, Italy
[2] Faculty of Engineering, Uninettuno University, 00186 Roma, Italy; livio.conti@uninettunouniversity.net
* Correspondence: baingio@uniss.it; Tel.: +39-3926315770

Received: 7 January 2020; Accepted: 13 January 2020; Published: 17 January 2020

Abstract: In this work, we discussed and counter-commented van der Helm's comments on our previous paper (Pinna and Conti, *Brain Sci.*, **2019**, *9*, 149), where we demonstrated unique and relevant visual properties imparted by contrast polarity in eliciting amodal completion. The main question we addressed was: "What is the role of shape formation and perceptual organization in inducing amodal completion?" To answer this question, novel stimuli were studied through Gestalt experimental phenomenology. The results demonstrated the domination of the contrast polarity against good continuation, T-junctions, and regularity. Moreover, the limiting conditions explored revealed a new kind of junction next to the T- and Y-junctions, respectively responsible for amodal completion and tessellation. We called them I-junctions. The results were theoretically discussed in relation to the previous approaches and in the light of the phenomenal salience imparted by contrast polarity. In counter-commenting van der Helm's comments we went into detail of his critiques and rejected all of them point-by-point. We proceeded by summarizing hypotheses and discussion of the previous work, then commenting on each critique through old and new phenomena and clarifying the meaning of our previous conclusions.

Keywords: amodal completion; contrast polarity; simplicity principle; likelihood principle; Bayes' framework

---

1. Introduction

Van der Helm [1] commented in a surprisingly polemical way on a recent paper of ours [2]. In this work, we answer his comments by proceeding in an objective manner on the basis of what was truly written and demonstrated in that paper on the basis of mere phenomenal evidence. In addition to the previous conditions, several others will be shown to make even more clear to the reader our previous purposes and conclusions, as well as the following counter-comments.

Our reply will be organized into sections based on van der Helm's comments. However, primarily for the sake of clarity, a short summary of Pinna and Conti [2] is required. This is useful to correctly inform readers about our previous purposes and conclusions necessary for a better understanding of our responses.

2. Pinna and Conti (2019) in Short

In this work, we demonstrated unique and relevant visual properties imparted by contrast polarity in perceptual organization and, more particularly, in eliciting amodal completion, where T-junction, good continuation, and closure are considered as the main factors involved.

Some of the most relevant explanations of amodal completion based on Helmholtz's likelihood principle [3] and Gregory's "unconscious inference" [4] were discussed. In short, the amodal object is similar to a perceptual hypothesis postulated to explain the unlikely gaps within the pattern of stimuli and the one that most likely produces the sensory stimulation (cf. the so-called avoidance-of-coincidences principle [5,6]). More recently these approaches have been reconsidered in terms of probabilistic Bayesian inference, applied successfully in many classical amodal conditions.

Another approach discussed within the paper is the simplicity–Prägnanz principle of Gestalt psychologists, according to which the visual system is aimed at finding the simplest and most stable organization consistent with the sensory inputs.

These approaches focus on the shape completed behind the occluder by assuming amodal completion as the cause of the formation of the shape partially occluded. Therefore, the main interest is to account for how the occluded object is completed, what is the amodal shape, and how contours of partially visible fragments are relatable behind an occluder.

Differently, we adopted a complementary perspective, assuming amodal completion not as the cause but as the resulting effect. Therefore, the questions are now the following: What is the role of shape formation and perceptual organization in inducing amodal completion? What are the perceptual conditions that elicit the segregation of occluded and occluding objects and, finally, amodal completion? What is the role of the local contours, junctions, and termination attributes in eliciting the phenomenon of amodal completion?

Within this perspective, amodal completion was considered as a visual phenomenon not as a process, i.e., the final outcome of perceptual processes and grouping principles. Therefore, the contrast polarity has been used as the main grouping and ungrouping factor aimed to explore and test amodal completion as a visual phenomenon elicited by good continuation, T-junctions, closure, and regularity.

Together with traditional configurations, novel stimuli, which have been reduced more and more to extreme limiting conditions, were introduced.

Phenomenally, we showed that contrast polarity is effective in inducing amodal completion in conditions where, on the basis of the known principles and of the previous theoretical approaches, it is not expected and vice versa. In this way, amodal completion was annulled or disrupted in the stimuli where it was supposed to be effective. More particularly, contrast polarity has been able to elicit/disrupt amodal completion when pitted against or in favor of the following conditions reduced to limiting cases: (i) Classical patterns (Figures 7 and 8); (ii) Petter's effect and Petter's rule (Figures 9–11); (iii) tessellation with T-junctions replaced by Y-junctions (Figures 12–16); (iv) group of isolated figures arranged in a cross (Figures 17–19); and (v) a single shape (Figures 20–22).

Our results demonstrated the compelling domination of the contrast polarity against good continuation, T-junctions, closure, and regularity. Moreover, the limiting conditions explored revealed a new kind of junction next to the T- and Y-junctions, respectively responsible for amodal completion and tessellation. They were called I-junctions, since eliciting amodal continuation of contours behind a contour with the same orientation.

Under our conditions, contrast polarity was shown to operate locally, eliciting results that could be independent from global scale and also paradoxical. We suggested that these results weaken and challenge theoretical approaches based on notions like oneness, unitariness, symmetry, regularity, simplicity, likelihood, constraints, and past knowledge. Moreover, Helmholtz's likelihood principle, simplicity/Prägnanz, and Bayes' inference are clearly questioned since they are supposed to operate especially at a global and holistic level of vision.

We proposed an alternative explanation of the specific outcomes based on the phenomenal dynamics made explicit by contrast polarity. First of all, the contrast polarity was perceived like a barrier analogous to, but stronger than, the one created by T-junctions. In other terms, the phenomenal salience, elicited by the highest luminance contrast going from black to white on a gray background, triggers a process of object segregation and unilateral belongingness of the boundaries.

The dominance of contrast polarity over good continuation, closure, regularity, and T-junctions is related to its phenomenal salience and highlighting effect. The same argument can account for the emergence of amodal continuation on I-junction, groups of isolated figures, and single shapes. Moreover, the imparted salience can disrupt, both locally and globally, arrangements of figures or can alternately rearrange the elements according to their similarity/dissimilarity. The highlighting strength of contrast polarity determines even the grouping effectiveness against the global and holistic rules and factors.

In favor of the basic and essential role of the phenomenal salience, we invoked deceiving strategies used in nature by most living organisms. Flowers, birds, and fishes use colors and contrast polarity to attract, reject, show, and hide. We suggest that the phenomenal salience strongly improves the biological fitness of living organisms and, therefore, the capability of an individual of a certain genotype to reproduce and, thus, to propagate an individual's genes within the genes of the next generation.

The phenomenal salience is a basic requirement also in human beings, in the way we dress, invent fashion and design, in the way we use the maquillage, and in the existence of the maquillage itself.

The strength of phenomenal salience imparted by contrast polarity enables the full independence from local or global organizations and top-down or bottom-up dynamics. It eludes all these categories since it can play in favor or against each of them, as demonstrated in our stimuli. It represents a true challenge for the theories discussed here, which cannot easily incorporate it (e.g., as a prior) without losing explanatory power somewhere else. Inside these arguments, it is important to underline that phenomenal salience is a perceptual attribute not restricted to contrast polarity, but it can also be triggered by color, shape, motion, and every other visual property. Among them, contrast polarity is one of the most powerful.

Given the significance of this attribute, further experimental studies based on phenomenological, psychophysical, and neurophysiological techniques are required to measure the strength of the phenomenal salience imparted by contrast polarity against other attributes involved in amodal completion. Further studies can shed light on the role of contrast polarity as a general tool useful in testing the range of scientific effectiveness of visual theories, approaches, and models.

This summary is taken on purpose almost verbatim from the final Discussion section of Pinna and Conti.

To make clear these hypotheses and, above all, to demonstrate the partial and incorrect interpretations and judgments of van del Helm, it is necessary to show the way we operate with reversed contrast. This is a novel way that cannot be reduced to the examples reported by van der Helm. This is the topic of the next section.

3. Phenomenology of Contrast Polarity

Van der Helm wrote: "The role of contrast polarity in human visual perception is a long-standing research topic (see, e.g., [7–10]). As a consequence, the phenomena presented by Pinna and Conti are not really as novel or surprising as they suggested them to be, but they are indeed illustrative of the effects of, in particular, contrast polarity reversals. For example, Figure 1 shows, in the style of Pinna and Conti, a stimulus in which such a reversal triggers a substantial change in the way in which it is perceptually organized."

As mentioned in the previous section, contrast polarity has been used in our paper as a grouping/ungrouping factor aimed to explore and test amodal completion as a visual phenomenon elicited by good continuation, T-junctions, closure, and regularity. It is no coincidence that the title of our work is: "The Limiting Case of Amodal Completion: The Phenomenal Salience and the Role of Contrast Polarity." A first difference in relation to the works [3–7], mentioned by van der Helms, is in the purposes but, more importantly, in the pattern of stimuli. In fact, while it is true that the quoted papers studied the role of contrast polarity, none of them used this factor the way we did, as can be easily demonstrated through their reading and especially by comparing their conditions with ours.

This entails that, differently from van der Helm's words—"The role of contrast polarity in human visual perception is a long-standing research topic (see, e.g., [7–10]). As a consequence, the phenomena presented by Pinna and Conti are not really as novel or surprising as they suggested them to be . . . .")—the conclusion, "As a consequence," cannot be a correct logical implication. The fact that other scientists studied "the role of contrast polarity in human visual perception" does not imply that our conditions are not new. In fact, although the first proposition is true, the second is not necessarily true and false, as we are going to demonstrate. As a consequence, the resulting implication is false according to the associated logic truth table.

This is not enough for a proper defense. We will now demonstrate the novelty of our conditions by showing some of them with further supporting examples. We will proceed going from known and trivial cases, as the one shown by van der Helm, to more and more interesting cases and demonstrations.

We start again from van der Helm's words already quoted: " ... For example, Figure 1 shows, in the style of Pinna and Conti, a stimulus in which such a reversal triggers a substantial change in the way in which it is perceptually organized."

The example reported is only partially in our style. To be more precise, it is just the starting condition that in our paper became much more complex figure after figure. Van der Helm played here with perceptual grouping by pitting similarity against good continuation. This is a classical procedure used by Gestalt psychologists. A similar basic condition is illustrated in the following stimulus (cf. Figure 1).

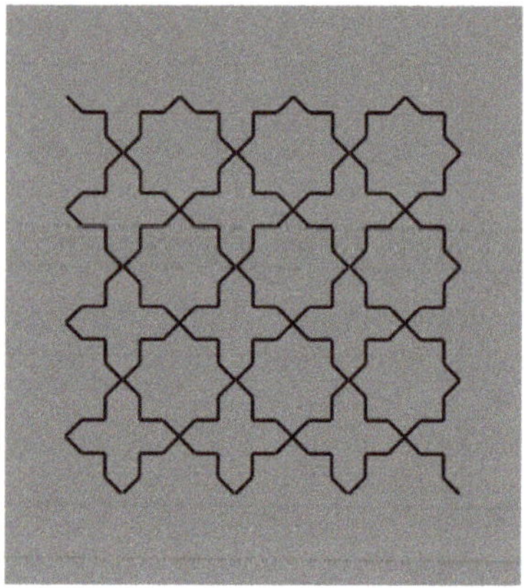

Figure 1. Crosses or stars? Alternated figure-ground segregation.

Here, we do not perceive individual segments, unconnected and oriented in different directions, but we see two main alternated and complementary shapes: Crosses and eight-pointed stars. By perceiving the crosses, the stars are invisible and vice versa. This is related to the unilateral belongingness of the boundaries. The two main results of Figure 1 are reversible and they can be easily switched by changing the focus of attention or just by moving the gaze in different locations of the stimulus pattern.

By introducing the grouping principle of similarity on the basis of the reversed luminance contrast, the salience of the crosses is now enhanced to the detriment of the stars (Figure 2, left). The opposite outcome is perceived in Figure 2, right.

The story becomes more interesting in Figure 3, where, reversing the contrast of only one of the two objects is sufficient to highlight and, thus, elicit the emergence of all the other similar objects. The accentuation of one object (e.g., one cross) spreads to all the others (all the remaining crosses). This figure suggests that reversed contrast could act as an accentuation principle [11–16] and not just as similarity.

The strong connection between figure-ground segregation and grouping is clearly shown in Figure 4, where the closure and proximity principles induce a clear object segmentation: Stars on the left and crosses on the right.

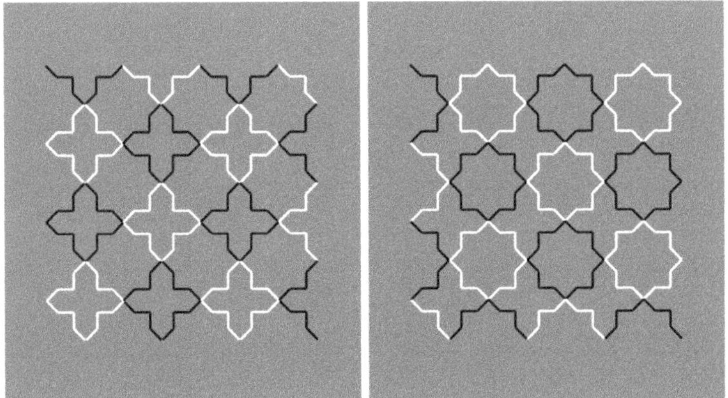

Figure 2. Crosses and stars due to the similarity principle of contrast polarity.

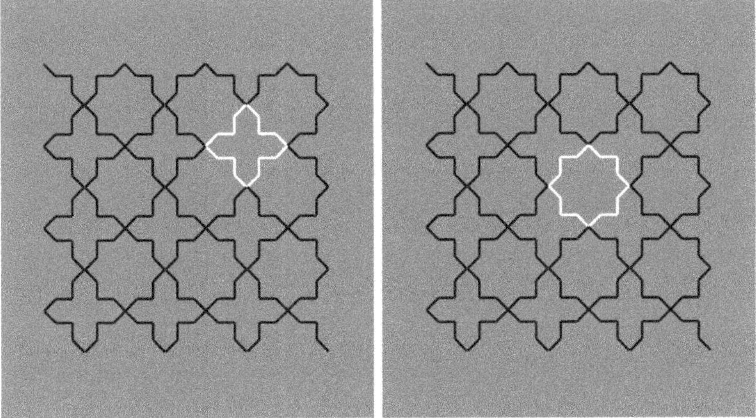

Figure 3. Crosses and stars spreading and filling in.

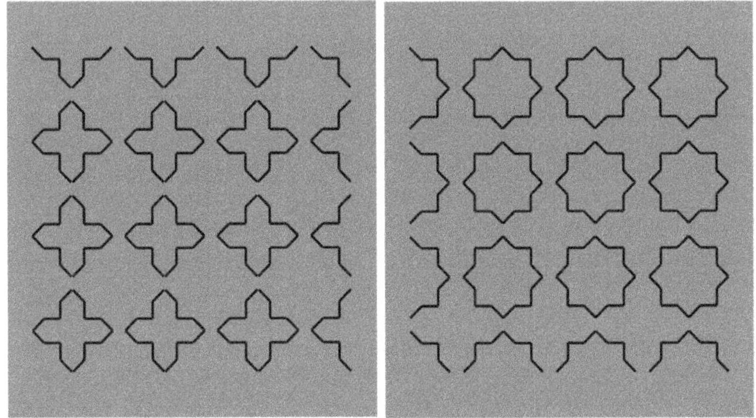

Figure 4. Crosses and stars elicited by the closure and proximity principles.

In Figure 5, contrast polarity enables the emergence of the complementary regions or, at least, it weakens the results of Figure 4.

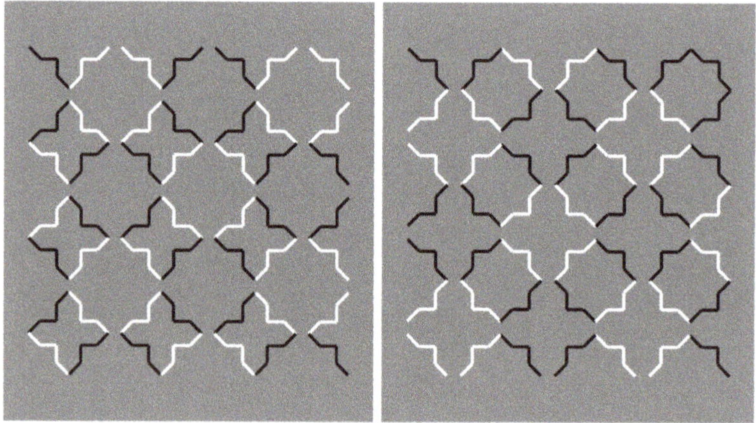

Figure 5. Similarity by contrast polarity pitted against closure and proximity principles.

Figure 6 shows a further and more interesting example, already reported within the paper.

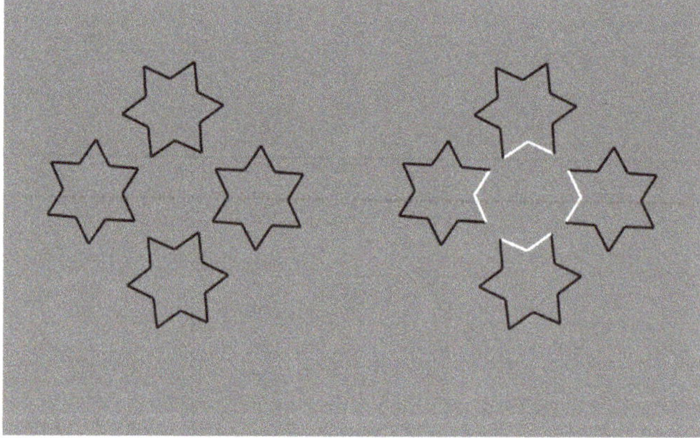

Figure 6. Similarity by contrast polarity pitted against closure and proximity principles.

Further conditions are illustrated in Figure 7. Here, by overlapping six stars, the grouping of the components, recombined by contrast polarity, becomes even more complex and unexpected.

Under these conditions, the regularity and homogeneity of the resulting object is impaired.

The strength of contrast polarity in terms of salience against other principles can be further perceived and appreciated in Figures 8 and 9.

Note that we are now introducing the term "salience," as a phenomenal attribute imparting a strong highlighting effect. Under these conditions, contrast polarity is used as a special case of the similarity principle, i.e., one of the many possible kinds of similarity, but in the next figures, due to the salience effect, it can be considered as a special case of the accentuation principle.

It is worth emphasizing that among the similarities, the reversed contrast is, in our opinion, one of the strongest and likely the strongest factor, and, as it happens in Figure 9, this is due to its highlighting effect that immediately jumps to the eye. Here, starting from two nested eight-pointed stars, namely, from two simple, singular, symmetrical, and Prägnant figures, we can highlight and then pop out new configurations.

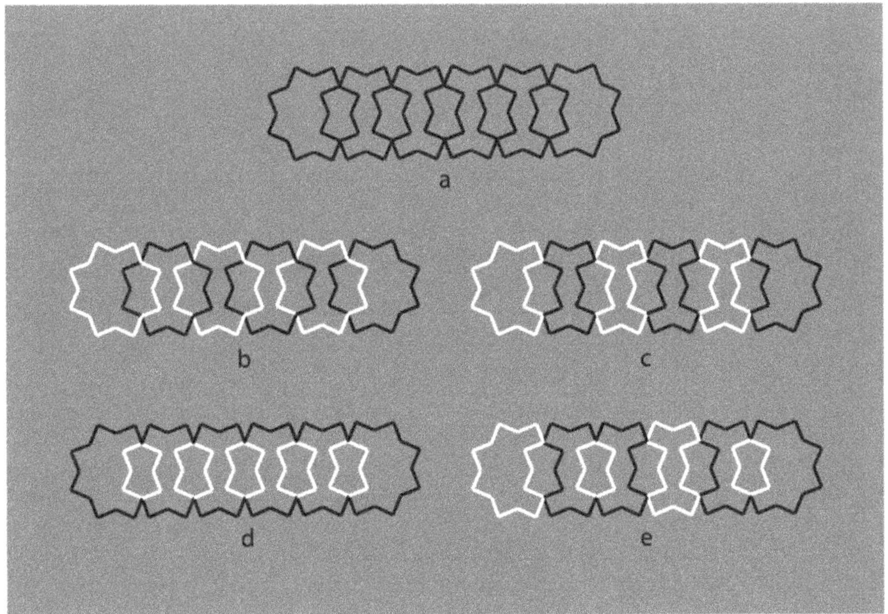

Figure 7. New complex results by overlapping six eight-pointed stars.

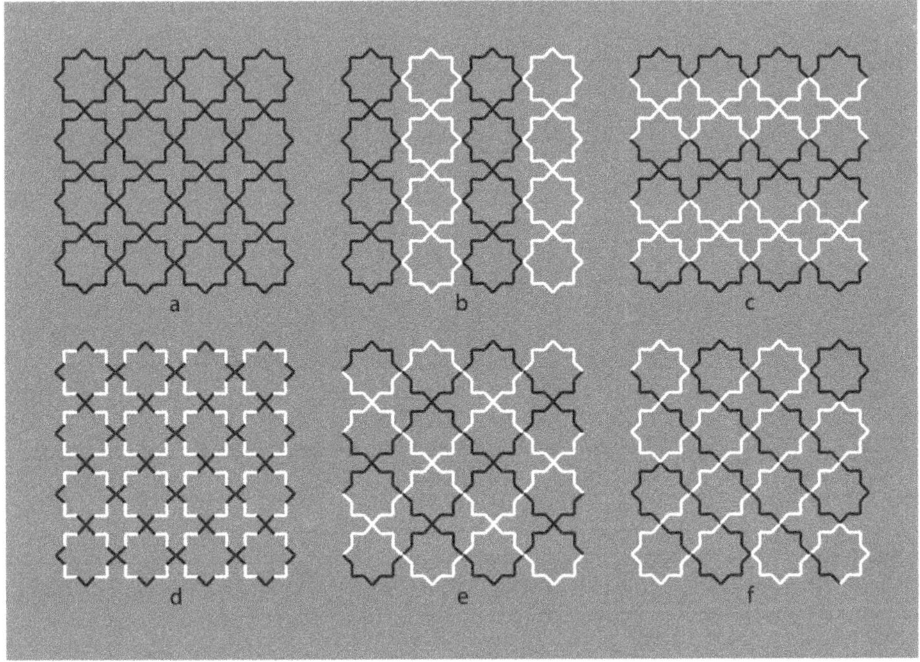

Figure 8. New regular and simpler results elicited by contrast polarity.

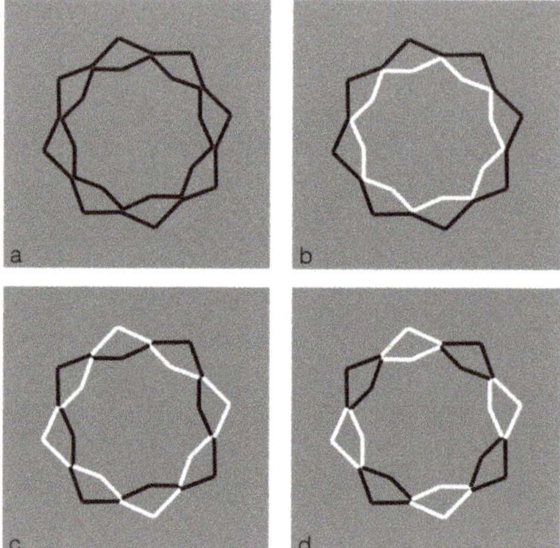

Figure 9. Different results on the basis of contrast polarity.

Figure 10 shows Figure 9a that becomes a sort of intertwined flower-like spiral without clean and univocal external boundaries.

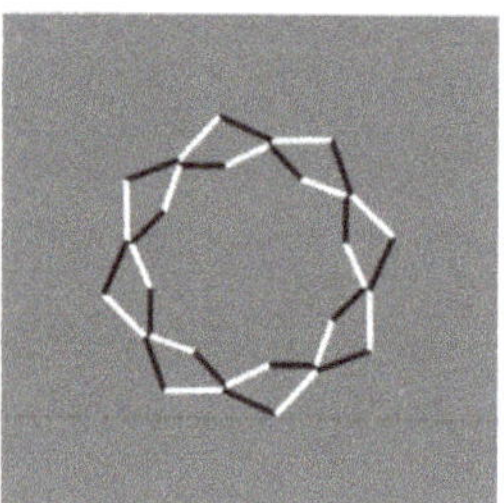

Figure 10. An intertwined flower-like spiral.

Regularity and likelihood of the previous figures become irregular and unlikely in Figure 11 and, more efficiently, in Figure 12.

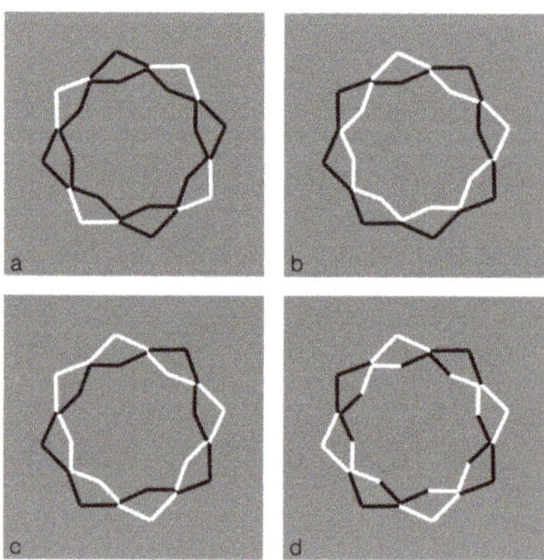

Figure 11. Iregular shapes from contrast polarity.

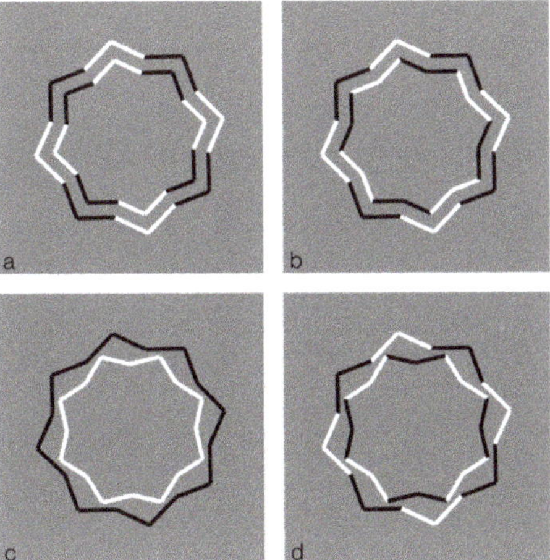

Figure 12. Regular and irregular shapes from reversed contrast.

From these examples, contrast polarity is not used as a similarity principle but more and more to accentuate and highlight new elements previously invisible and camouflaged. This idea, already discussed in our paper, will be developed in the next figures starting from the following variations of Necker cube (cf. Figure 13).

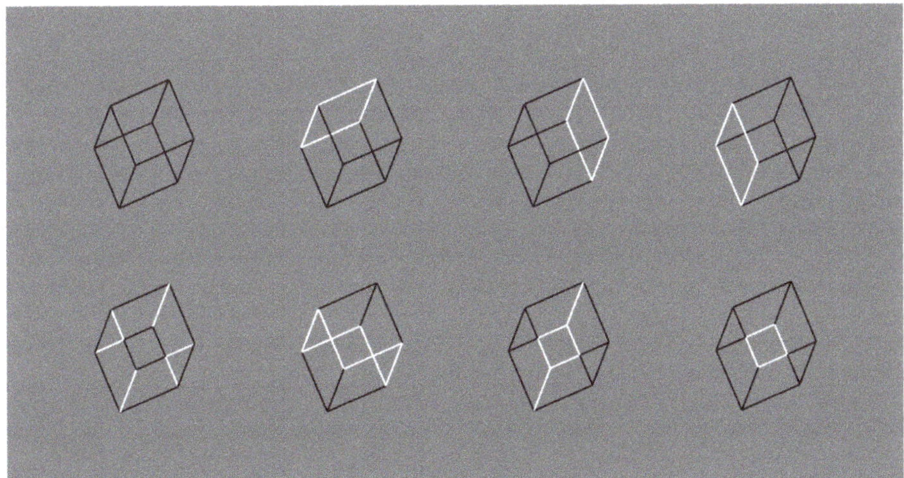

Figure 13. Different views of the Necker cube (first row) and weakened, disrupted, camouflaged, and invisible cube (second row).

What is worthwhile to be considered in this figure is the set of cubes of the first row, where highlighting one view or the other, by means of reversed contrast, the same cube seems to be differently oriented in 3D space. The reversed contrast is now more related to accentuation principle than to similarity.

Given the salience effect, we finally explore limiting conditions, more directly related to simplicity and likelihood approaches. These conditions represent a further level of application of the reversed contrast, no more within multiple contours highlighting, disrupting, or camouflaging one or another possible configuration, but within a unique closed shape, more simple as possible, and where the possibility to perceive further inset subcomponent is very unlikely. These cases allow to test more appropriately simplicity and likelihood. Some of these conditions are illustrated in Figure 14.

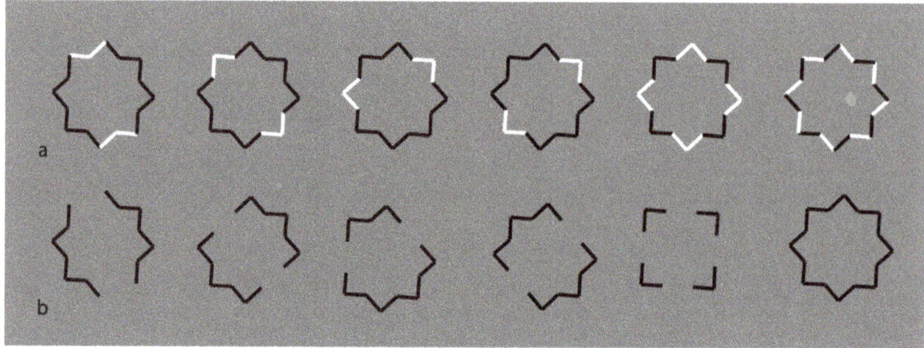

Figure 14. The black boundaries of a regular eight-pointed star discontinued by reversing the contrast polarity.

Now, the black boundaries of a regular eight-pointed star have been discontinued by reversing the contrast polarity in different portions of each figure. In terms of grouping, the closure principle is mainly responsible for the perception of the eight-pointed star. Phenomenally, the uniqueness and wholeness of each star illustrated appear weakened or broken down. Each shape of the first row partially splits into two black and white adjacent components that are seen as belonging to two different objects.

The most extensive black components reveal more easily the whole shape that, given the ungrouped white sides, do not appear as eight-pointed stars, but as the related shapes illustrated in Figure 14b, with inner gaps interrupting the boundary continuation. The removal of the white components improves the grouping of the black segments on the basis of the closure and good continuation principles. The missing sides do not affect the whole shape but appear as gaps favoring the good continuation of the modal sides and the amodal wholeness [12]. Given the black and white alternation of the sides, only the last condition of the first row appears as analogous to the star of the last condition of Figure 14b.

Further examples are illustrated in Figure 15. Again, the reverse contrast breaks the oneness and the unitariness of the eight-pointed star and influences significantly its shape. New organizations emerge as follows: A concave polygonal shape rather than a star (Figure 15a); two rotated and perpendicular intersecting square shapes with illusory curved sides (Figure 15b); irregular shapes different from stars in Figure 15c–g. Figure 15h is the control.

Figure 15a,b display something that is worth noticing, namely the emergence of the concavity and of the convexity within the same geometrical star. As a matter of fact, the two figures are perceived as stars. However, the first appears as a sort of 'eight-concave star' or 'eight-indented star.'

The breaking of the uniqueness and unitariness is very effective even though good continuation, closure, and Prägnanz together favor the grouping. The dissimilarity imparted by the reversed contrast split the figure. The question is: Why are these irregular and parceled solutions preferred to a simpler result like 'a star with black and white edges?' With this solution, both the whole shape and the local changes would have been preserved and saved. Even if this ideal solution is possible, it does not prevail.

These outcomes clearly challenge the theoretical assumptions of oneness, unitariness, symmetry, regularity, simplicity, minimization of description length, likelihood, and previous knowledge.

These limiting conditions can be even more pushed over by drastically reducing the role of the discontinuities due to the reversed polarity and, therefore, by improving the good continuation and the similarity of remaining parts of the figure. This is the case of the polygons illustrated in Figure 16, where the perception of a polygon is the only possibility with the maximum likelihood.

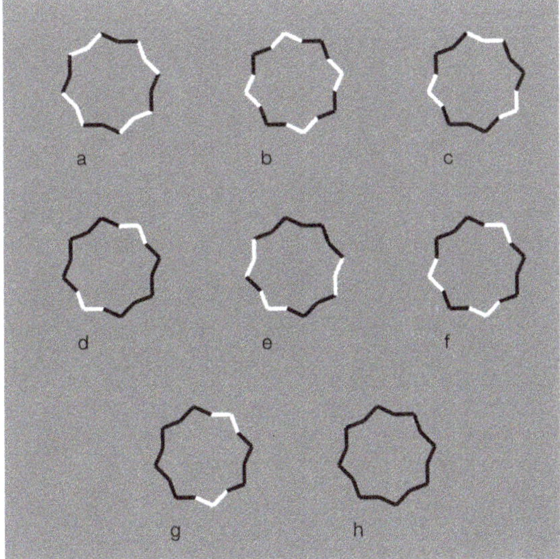

Figure 15. Further examples of the role played by contrast polarity in breaking the unitariness of a star.

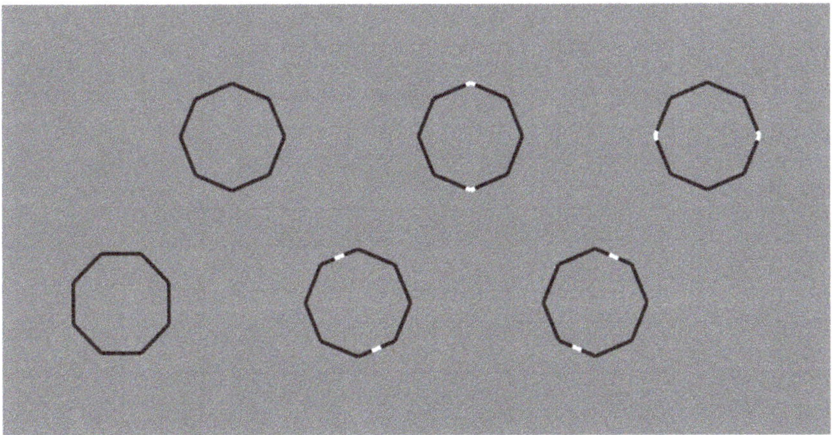

Figure 16. Octagons apparently different due to the accentuation imparted by contrast polarity.

In Figure 16 top-center, a regular octagon is perceived with the vertices oriented along the main directions of space (vertical and horizontal). By introducing achromatic discontinuities on two vertices placed along the vertical or on the horizontal axes of the polygon (second and third figure in the first row), the same kind of oriented regular octagon is seen. However, by introducing similar discontinuities in two opposite sides placed along an oblique axes (second and third figure in the second row), the polygons are now perceived as equal and, at the same time, different: Their shapes appear more similar to the first octagon of the second row, i.e., with the sides, rather than the vertices, oriented along the main directions of space. Moreover, the second and third octagons of the second row appear as oriented on opposite oblique directions with respect to the vertical axes.

Similar results can be produced with just a single white discontinuity (see Figure 17).

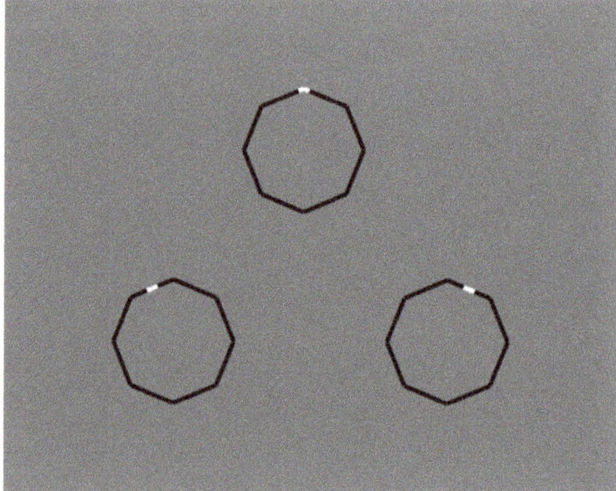

Figure 17. The octagons appear different due to a single white accentuation imparted by the contrast polarity.

Unlike the stars of Figures 14 and 15, the wholeness of the polygons is not broken down or disrupted. This effect mainly occurs due to the significant size reduction of the white components that let the good continuation and, hence, the unitariness and the wholeness of the boundaries at work. Phenomenally, the white discontinuities, placed on the vertices (angles) and sides, highlight and accentuate within the same shape different components (angles and sides) and object attributes.

The white dashes within the polygons behave phenomenally like accents inducing the pop out of one or another geometrical basic component of the polygon, i.e., angle or side. The white dashes do not represent the source of the disruption of the object, as shown in the previous figures, but behave like accents of a specific attribute of the figure that is ipso facto highlighted. This accent determines a change of the whole object, not only of its orientation or tilt, but of one of the two attributes. In other terms, the accent changes the nature and the geometry of the object. The accentuated pointedness and sidedness create two different polygons (same but different), one more pointed and the other flatter.

The accentuation induced by contrast polarity can also disrupt the regularity of the octagons as illustrated in Figure 18.

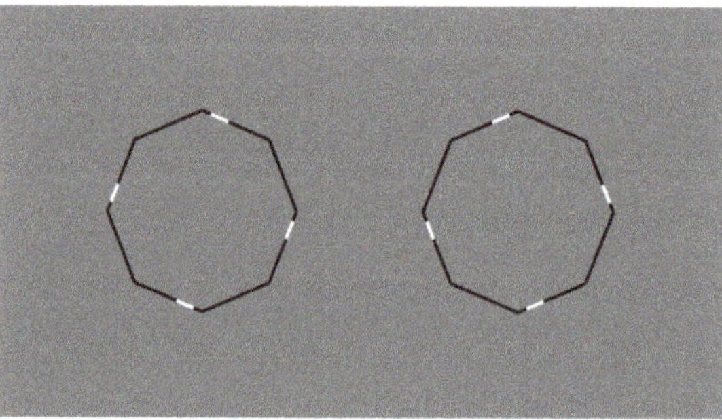

Figure 18. The octagons appear irregular and oriented in opposite directions.

Counter Comments: Reversed Contrast and Amodal Completion

Differently from van der Helm's judgment, the stimuli presented in this section demonstrate the novelty of the way contrast polarity was used.

It is surprising that van der Helm based his critiques only on the example produced by himself and not on our stimuli. This is unfair, and in our view, scientifically inappropriate, to quote the same words used by van der Helm. It would have been scientifically more correct to comment in detail on each of our figures.

Further interesting conditions for the discussion are shown in Figures 19–21. In Figure 19, the reversed contrast is pitted in favor or against T-junctions, good continuation, closure, simplicity, and likelihood. A detailed description of this set of stimuli can be found in our paper.

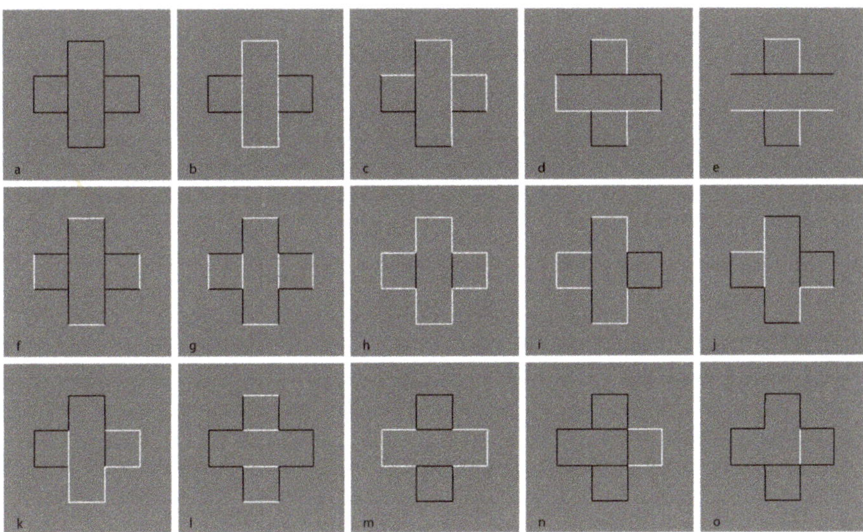

Figure 19. Conditions demonstrating the dominance of the contrast polarity over good continuation, T-junctions, regularity, simplicity, and likelihood.

Unfortunately, van der Helm did not discuss these stimuli that represent the core of our work. Moreover, he did not notice the novelty of the I-junctions, more clearly shown in Figure 20 (Figure 21 is a control), where the continuation of contours is behind a contour with the same orientation. In other terms, two contours, although showing the same orientation, are perceived as two different contours where one appears to complete amodally behind the other. The effect of the I-junctions occurs against T-junctions, regularity, good continuation, closure, simplicity, likelihood. For a detailed description of this novel outcome see Pinna & Conti [2].

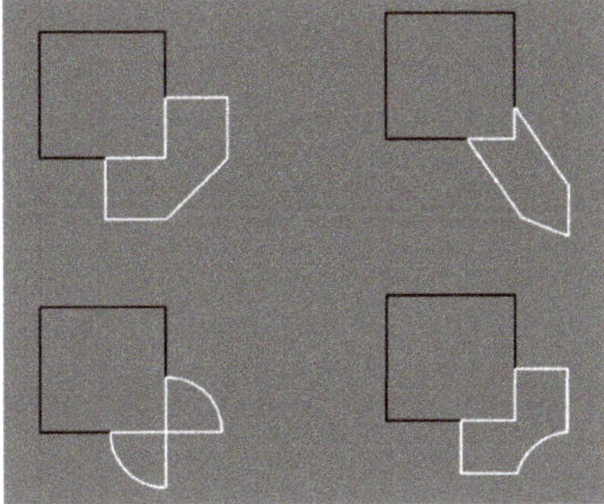

Figure 20. The contrast polarity reverses the amodal completion: The square is now perceived as partially occluded by the asymmetrical objects (for a control see Figure 21).

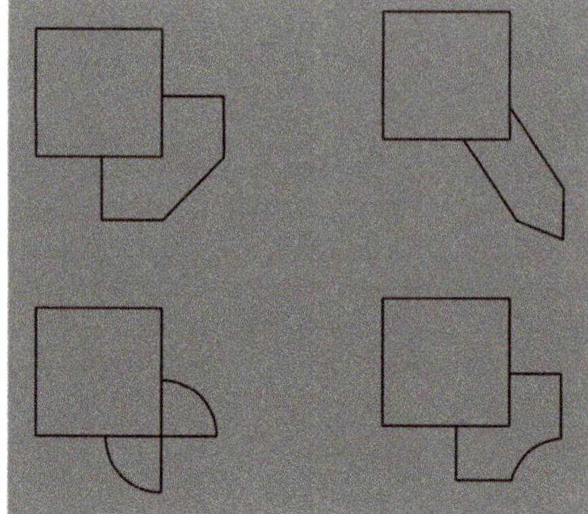

Figure 21. A control for Figure 20. The figures, partially occluded by the square, complete amodally as asymmetrical shapes.

4. On Local vs. Global Assumption: Likelihood and Simplicity

Van der Helm judged and labelled three sentences of our paper as incorrect assumptions. The sentences are the following:

i. The salience and visibility, derived by the largest amplitude of luminance dissimilarity imparted by contrast polarity, precedes any holist or likelihood organization due to simplicity/Prägnanz and Bayes' inference.

ii. Contrast polarity was shown to operate locally, eliciting results that could be independent from any global scale and that could also be paradoxical. These results weaken and challenge theoretical approaches based on notions like oneness, unitariness, symmetry, regularity, simplicity, likelihood, priors, constraints, and past knowledge. Therefore, Helmholtz's likelihood principle, simplicity/Prägnanz and Bayes' inference were clearly questioned since they are supposed to operate especially at a global and holistic level of vision.

iii. The highlighting strength of contrast polarity determines even the grouping effectiveness against the global and holistic rules and factors expected by Helmholtz's likelihood principle, simplicity/Prägnanz and Bayes' inference.

A simple response to this judgement is that the three sentences are not assumptions but just evidence based on our results and on the phenomenology of the stimuli. On the contrary, we think that van der Helm's judgement is based on an incorrect assumption. In fact, he did not comment on the three sentences on the phenomenology of our effects, i.e., by showing and demonstrating that our results are not consistent with our conclusion. Instead, he assumed as incorrect our sentence on the base of the following results obtained by van Lier:

It is true that simplicity and likelihood approaches may aim to arrive at global stimulus interpretations, but a general objection against the above stance is that they (can) do so by including local factors as well. For instance, van Lier [16] presented a theoretically sound and empirically adequate simplicity model for the integration of global and local aspects in amodal completion.

We think that this is unfair and scientifically inappropriate, in his own words. In fact, he continues:

A methodological objection is that Pinna and Conti introduced contrast polarity changes in stimuli but pitted these against alleged simplicity and likelihood predictions for the unchanged stimuli. As I specify next, this is unfair, and in my view, scientifically inappropriate.

Starting from our purpose—according to which contrast polarity has been used as the main grouping and ungrouping factor aimed to explore and test amodal completion as a visual phenomenon elicited by good continuation, T-junctions, closure, and regularity—this is not unfair and it is not scientifically inappropriate since our predictions are not based on unchanged stimuli but, on the contrary, on changed stimuli. For example, in the case of Figure 21 of our paper, we stated: "The breaking of the uniqueness and unitariness is very effective even though good continuation, closure, and, even, Prägnanz together favor the grouping. The dissimilarity imparted by the reversed contrast splits the figure. The question is: why are these irregular and parceled solutions preferred to a simpler result like 'a star with black and white edges?' With this solution, both the whole shape and the local changes would have been preserved and saved. Even if this ideal solution is possible, it does not prevail."

A further van der Helm's critique states as follows:

> *Pinna and Conti wrote "[If] we do not consider the contrast polarity as a constraint or as a prior, Bayes' inference cannot easily explain these conditions" [2] (p. 12 of 32)—indeed, but why would we? Hence, they knowingly ignored the above flexibility and applied likelihood as if it is fundamentally blind to contrast polarity. Thereby, they missed the mark in their assessment of likelihood approaches.*

We did not ignore the flexibility of the Bayesian approach; in fact we wrote: "Bayes' inference cannot easily explain these conditions." We just reported the most immediate prediction that could be made and leave open the point. Nevertheless, van der Helm based his judgement on vague conjecture and this cannot be accepted. On the contrary, it would have been acceptable whether he had explained our stimuli according to Bayesian inference.

We do deny that this kind of explanation might be possible, we are not against Bayesian inference, but we just casted doubts about its effectiveness in the light of our stimuli.

In commenting on simplicity, van der Helm refers to Figure 2, introduced by himself, and draws conclusions on our work by stating: "Pinna and Conti knowingly ignored this and applied simplicity as if it is fundamentally blind to contrast polarity. Thereby, they missed the mark in their assessment of simplicity approaches". We cannot accept this gratuitous judgment and critique based on general hypothesis, far away from our work. We really do not understand this attitude. We would have preferred a direct comment to each of our figures. This is much more useful to improve theories, useful for science.

Anyway, we have never said that simplicity is blind to contrast polarity. We do not believe so. Our phenomenological background is based on the notion of simplicity. Our point was just focused on the conditions we studied. As a matter of fact, our results put to the test these important theories and approaches in line with Popper's falsificationism. This is scientifically appropriate.

5. On the Equivalence between Simplicity and Likelihood

Van der Helm quoted the unique sentence in our paper where he thinks we claim the equivalence between simplicity and likelihood: "[...] the visual object that minimizes the description length is the same one that maximizes the likelihood. In other terms, the most likely hypothesis about the perceptual organization is also the outcome with the shortest description of the stimulus pattern." [2] (p. 3 of 32). He also wrote: "This is an extraordinary claim".

As a first reply to this comment, we can confidently say that it was far from our purpose to demonstrate or discuss the supposed equivalence. Our purposes were very different as described in detail in the first section of the present work. Second, the supposed equivalence is just related and based on our conditions only without any further generalization. In our opinion, based on our results, simplicity and likelihood do not deliver different predictions. For van der Helm, it would have been easier to indicate possible differences to give a more effective contribution.

We cannot accept that van der Helm can attribute to us generalizations never mentioned. Again, the supposed equivalence is implicit and restricted to only our conditions. We did nothing to prove or disprove this equivalence being far from our purposes. Third, in any case, as acknowledged even by van der Helm, this equivalence is not an extraordinary claim. Other scientists (see below) are in favor of some sort of equivalence.

According to Helmholtz's likelihood principle, the sensory input is organized into the most probable distal object or event consistent with the sensory data (the proximal stimulus). This principle chooses the most likely true interpretation and assumes that the visual system is highly veridical in terms of the external world. From an evolutionary point of view, if the visual system were not veridical, it would probably not have survived during the evolution.

On the other hand, the simplicity principle does not experience these problems, because it does not aim specifically at veridicality. The simplicity and the likelihood principle are two competing theories [5,17–19] of perceptual organization and visual coding, which are difficult to settle because neither of the key elements were clearly defined. The general difference between the two is related to the fact that the visual system, in the case of the simplicity, obeys a more general principle of economy, while in the case of the likelihood, it obeys a general principle of probability.

Nevertheless, these two terms might be only apparently different or may be considered as two sides or two different ways of considering the same visual process. Mach [20] suggested that vision acts in conformity with the principle of economy, and, at the same time, in conformity with the principle of probability. Chater [21] demonstrated mathematically that these key elements can be unified and considered equivalent within the theory of Kolmogorov complexity [22–27].

Feldman [28–31] presented a simplicity approach, called minimal model theory, and, in agreement with Chater [21], suggested that the visual interpretation, whose description is of minimum length, is the one that most likely is also the most veridical. Usually, the tendency of choosing a visual object that minimizes the description length is the same as the tendency of choosing a hypothesis that maximizes the likelihood. In brief, the most likely hypothesis about perceptual organization is, at the same time, the objects supporting the shortest description of the stimulus.

Far from this controversy, the point is here that we firmly reject van der Helm's attribution of a supposed general equivalence between the two principles.

6. Conclusions

We discussed and counter-commented van der Helm's [1] comments on Pinna and Conti [2] going into detail of his critiques and rejecting all of them point-by-point. We proceeded by summarizing hypotheses and discussion of the previous work, then commenting on each critique through old and new phenomena and clarifying the meaning of our previous conclusions. It is a pity that van der Helm's comments were not directly related to our stimuli. This would have stimulated a more effective discussion and possibly some new ideas. Anyway, we hope that this exchange of views can stimulate other scientists to further develop theories and phenomena related to amodal completion and reversed contrast. We are convinced about the importance of these intriguing objects, not fully explained yet, and that deserve to be further studied through different perspectives. Finally, we hope that Bayesian simulations can cast new light on the hypothesis presented in our paper and in this discussion.

Acknowledgments: Supported by "Fondo di Ateneo per la ricerca 2019", by Fondazione Banco di Sardegna (finanziato a valere sulle risorse del bando Fondazione di Sardegna—Annualità 2015) and Alexander von Humboldt Foundation.

Conflicts of Interest: The authors declare no conflict of interest. The funders had no role in the design of the study; in the collection, analyses, or interpretation of the data; in the writing of the manuscript; or in the decision to publish the results.

References

1. Van der Helm, P.A. Dubious Claims about Simplicity and Likelihood: Comment on Pinna and Conti (2019). *Brain Sci.* **2020**, *10*, 50.
2. Pinna, B.; Conti, L. The limiting case of amodal completion: The phenomenal salience and the role of contrast polarity. *Brain Sci.* **2019**, *9*, 149. [CrossRef] [PubMed]
3. Von Helmholtz, H. *Handbuch der Physiologischen Optik*; Leopold Voss: Leipzig, Germany, 1867.
4. Gregory, R.L. Perceptions as hypotheses. *Philos. Trans. R. Soc. Lond.* **1980**, *B290*, 181–197. [CrossRef]

5. Rock, I. *The Logic of Perception*; MIT Press: Cambridge, MA, USA, 1983.
6. Rock, I. Perception and knowledge. *Acta Psychol.* **1985**, *59*, 3–22. [CrossRef]
7. Bell, J.; Gheorghiu, E.; Hess, R.F.; Kingdom, F.A.A. Global shape processing involves a hierarchy of integration stages. *Vis. Res.* **2011**, *51*, 1760–1766. [CrossRef] [PubMed]
8. Elder, J.; Zucker, S. The effect of contour closure on the rapid discrimination of two-dimensional shapes. *Vis. Res.* **1993**, *33*, 981–991. [CrossRef]
9. Schira, M.M.; Spehar, B. Differential effect of contrast polarity reversals in closed squares and open L-junctions. *Front. Psychol. Percept. Sci.* **2011**, *2*, 47. [CrossRef] [PubMed]
10. Spehar, B. The role of contrast polarity in perceptual closure. *Vis. Res.* **2002**, *42*, 343–350. [CrossRef]
11. Su, Y.; He, Z.J.; Ooi, T.L. Surface completion affected by luminance contrast polarity and common motion. *J. Vis.* **2010**, *10*, 5. [CrossRef]
12. Pinna, B. New Gestalt principles of perceptual organization: An extension from grouping to shape and meaning. *Gestalt Theory* **2010**, *32*, 11.
13. Pinna, B. What Comes Before Psychophysics? The Problem of 'What We Perceive' and the Phenomenological Exploration of New Effects. *See. Perceiving* **2010**, *23*, 463–481. [CrossRef] [PubMed]
14. Pinna, B. What is the meaning of shape? *Gestalt Theory* **2012**, *33*, 383–422.
15. Pinna, B. Perceptual Organization of Shape, Color, Shade, and Lighting in Visual and Pictorial Objects. *i-Perception* **2012**, *3*, 257–281. [CrossRef] [PubMed]
16. Pinna, B. The Place of Meaning in Perception—Introduction. *Gestalt Theory* **2012**, *33*, 3221–3244.
17. Pinna, B.; Reeves, A.; Koenderink, J.; van Doorn, A.; Deiana, K. A new principle of figure-ground segregation: The accentuation. *Vis. Res.* **2018**, *143*, 9–25. [CrossRef]
18. Hatfield, G.; Epstein, W. The status of the minimum principle in the theoretical analysis of visual perception. *Psychol. Bull.* **1985**, *97*, 155–186. [CrossRef]
19. Pomerantz, J.R.; Kubovy, M. Theoretical approaches to perceptual organization: Simplicity and likelihood principles. In *Handbook of Perception and Human Performance, Volume II*; Boff, K.R., Kaufman, L., Thomas, J.P., Eds.; John Wiley & Sons, Inc.: New York, NY, USA, 1986.
20. Leeuwenberg, E.L.; Boselie, F. Against the likelihood principle in visual form perception. *Psychol. Rev.* **1988**, *95*, 485–491. [CrossRef]
21. Mach, E. *The Analysis of Sensations and the Relation of the Physical to the Psychical*; Dover: New York, NY, USA, 1959.
22. Chater, N. Reconciling simplicity and likelihood principles in perceptual organization. *Psychol. Rev.* **1996**, *103*, 566–581. [CrossRef]
23. Solomonoff, R.J. A formal theory of inductive inference, Part 1. *Inf. Control* **1964**, *7*, 1–22. [CrossRef]
24. Solomonoff, R.J. A formal theory of inductive inference, Part 1 and 2. *Inf. Control* **1964**, *7*, 224–254. [CrossRef]
25. Chaitin, G.J. On the length of programs for computing finite binary sequences: Statistical considerations. *J. Assoc. Comput. Mach.* **1969**, *16*, 145–159. [CrossRef]
26. Kolmogorov, A.N. Three approaches to the quantitative definition of information. *Probl. Inf. Transm.* **1965**, *1*, 1–7. [CrossRef]
27. Rissanen, J.J. Modelling by the shortest data description. *Automatica* **1978**, *14*, 465–471. [CrossRef]
28. Li, M.; Vitànyi, P. *An Introduction to Kolmogorov Complexity and Its Applications*, 2nd ed.; Springer: New York, NY, USA, 1997.
29. Feldman, J. Bayesian contour integration. *Percept. Psychophys.* **2001**, *63*, 1171–1182. [CrossRef] [PubMed]
30. Feldman, J. Formation of visual "objects" in the early computation of spatial relations. *Percept. Psychophys.* **2007**, *69*, 816–827. [CrossRef] [PubMed]
31. Feldman, J.; Singh, M. Bayesian estimation of the shape skeleton. *Proc. Natl. Acad. Sci. USA* **2006**, *103*, 18014–18019. [CrossRef]

Article

Imagery-Mediated Verbal Learning Depends on Vividness–Familiarity Interactions: The Possible Role of Dualistic Resting State Network Activity Interference

Etienne Lefebvre and Amedeo D'Angiulli *

Neuroscience of Imagination Cognition and Emotion Research (NICER) Lab, Neuroscience Department, Carleton University, Ottawa, ON K1S 5B6, Canada; EtienneLefebvre@cmail.carleton.ca
* Correspondence: amedeo.dangiulli@carleton.ca

Received: 2 May 2019; Accepted: 17 June 2019; Published: 18 June 2019

Abstract: Using secondary database analysis, we tested whether the (implicit) familiarity of eliciting noun-cues and the (explicit) vividness of corresponding imagery exerted additive or interactive influences on verbal learning, as measured by the probability of incidental noun recall and image latency times (RTs). Noun-cues with incongruent levels of vividness and familiarity (high/low; low/high, respectively) at encoding were subsequently associated at retrieval with the lowest recall probabilities, while noun-cues related with congruent levels (high/high; low/low) were associated with higher recall probabilities. RTs in the high vividness and high familiarity grouping were significantly faster than all other subsets (low/low, low/high, high/low) which did not significantly differ among each other. The findings contradict: (1) associative theories predicting positive monotonic relationships between memory strength and learning; and (2) non-monotonic plasticity hypothesis (NMPH), aiming at generalizing the non-monotonic relationship between a neuron's excitation level and its synaptic strength to broad neural networks. We propose a dualistic neuropsychological model of memory consolidation that mimics the global activity in two large resting-state networks (RSNs), the default mode network (DMN) and the task-positive-network (TPN). Based on this model, we suggest that incongruence and congruence between vividness and familiarity reflect, respectively, competition and synergy between DMN and TPN activity. We argue that competition or synergy between these RSNs at the time of stimulus encoding disproportionately influences long term semantic memory consolidation in healthy controls. These findings could assist in developing neurophenomenological markers of core memory deficits currently hypothesized to be shared across multiple psychopathological conditions.

Keywords: DMN; TPN; vividness; familiarity; memory; mental imagery

---

1. Introduction

Consider the following scenario: You forgot the password for your online banking account. Fortunately, you left yourself a password hint: "childhood telephone number". You used the number for many years. Yet, even though it felt so familiar, you could recall it only after experiencing a vivid image of yourself dialing it on your old rotary dial phone. This vignette simplifies the vivid incidental recall (VIR) effect. This effect has been replicated in experiments where participants are tasked with generating the visual mental images corresponding to a list of cuing object-nouns, and, after an intervening task or rest delay, they are surprised with the request to recall the words. The cues which correspond to relatively higher subjectively vivid images generally yield a higher probability of being successfully retrieved during unexpected recall [1–4].

Yet, the neurocognitive processes or variables underpinning effects, such as VIR, have not been addressed directly in terms of underlying memory mechanisms. A possible link can be made with dual process theories of recognition memory, according to which retrieval can take two forms: a recollection of contextual and qualitative details, or a subjective feeling of familiarity about the stimulus in question [5,6]. For stimuli to be recalled during an incidental recall task, the memory of the stimuli must be retrieved with its associated qualitative details, and therefore, must rely on the recollection process of recognition memory. On this account, VIR could occur because more vivid mental imagery strengthens the process of recollection, resulting in higher scores in incidental recall tasks. However, these theories also imply that another key factor should be involved in the vivid incidental recall effect, namely, previous and repeated exposure to the given stimulus, in this case, the familiarity associated with word cuing mental imagery and recall [7]. Still, it is unclear whether a level of familiarity strength similar (or congruent) to that of the vividness in the underlying representation may increase or reduce the success of its recollection, thereby modulating the extent of VIR effects. In other words: Are two different types of relatively weak encodings better than a single relatively strong one for consolidating the same memory? To rephrase in terms of our opening anecdotal vignette: Would we still be able to recall our password hint successfully if we had a vivid image of dialing it, but the number itself lost its familiarity? Would this incongruence between the level of familiarity and vividness hinder recollection? This question is the focus of this paper.

According to associative theories, such as Paivio's dual coding theory, two memory representations should have cooperative effects on the probability of accurate incidental retrieval [8]. That is, the relative strengths of two memory representations should be additive and should always yield a higher probability of reinstating recollection of the original stimulus during incidental recall. Accordingly, having only one strong encoding should be equivalent to having two different weak ones, and two strong encodings will always be better than two weak ones. Hence, the success of incidental retrieval should increase monotonically with the strength of the memory representation.

Most recently, however, accounts based on the neurobiology of learning and memory challenged associative explanations. In particular, the non-monotonic plasticity hypothesis (NMPH) holds that learning effects (recall accuracy) vary as a non-monotonic function of the amount of excitation associated with competing memory representations [9]. Specifically, NMPH predicts that strong learning effects result from overwhelmingly high levels of excitation in one of the two competing representations; poor learning effects result from moderate excitation in both representations, while no learning effects result from low levels of excitation overall [10]. That is, two strong encodings or a strong single one will always better than two weak ones, but two moderately intense encodings will result in the worst learning outcome. As a result, the neural strength of memory representations should be non-monotonically related to the probability of accurate incidental retrieval.

While the NMPH proposes a theory of memory formation and learning, it is first and foremost a neurobiological theory that is rooted at processes taking place at the level of the neuron. It is still unclear whether theories of synaptic strengthening, such as the non-monotonic plasticity hypothesis, are useful frameworks to utilize when modeling higher order phenomenon, such as memory consolidation and retrieval. Recent work using large functional brain networks have identified clear functional and spatial differential specialization for both memory consolidation and memory retrieval processes. More importantly, clear differences in neural specializations are observed when assessing the influence of the accuracy of memory retrieval (performance) as compared to the cortical activation observed at the time of encoding and retrieval [11,12].

A related, known challenge for theories which attempt to reduce the content of recollections to relatively simple neural mechanisms is that they do not adequately describe the relationships between phenomenal consciousness, for example, in the present context, what seems phenomenally available in imagery (i.e., vivid), and what is ultimately accessed in the underlying memory representations (see [13] for the general argument). This is indicative that building robust and replicable models of memory processes at the level of large functional brain networks can provide insightful analysis that may be able to bridge the gap between the phenomenal, the behavioral, and the neural.

Over the last decade, large functional brain networks have been reliably identified across a variety of independent cognitive states using both functional magnetic resonance imaging (fMRI) and Electroencephalography (EEG). These large functional brain networks, coined resting-state networks (RSNs), have been shown to be accurate biomarkers for different states of consciousness, and more importantly accumulating evidence suggests that abnormal RSN activation patterns underly numerous psychopathologies [14,15]. A growing consensus suggests that functional brain activity can be divided into two anti-correlated RSNs: namely, the default mode network (DMN) and the task-positive-network (TPN) [16]. The DMN is generally referred to as a baseline state or a mind wandering state that involves dynamic activity between multiple cortical and subcortical areas that principally involves the medial temporal subsystems (hippocampus, parahippocampal cortex, and retrosplenial cortex), and the dorsal medial subsystems (dmPFC, TPF, and the temporal pole) [17]. Functionally, the DMN is thought to be responsible for the regulation of emotional processing, self-referential mental activity, and the recollection of prior experiences [18]. The TPN is generally considered to reflect externally mediated cognition that involves activity in a number of subsystems, including the dorsal attention network (intraparietal sulcus, sections of the precentral and frontal sulcus, and middle frontal gyrus), the posterior visual network (retinotopic occipital cortex and the temporal-occipital region), the auditory–phonological network (bilateral superior temporal cortex), and the motor network (regions of the precentral, postcentral, and medial frontal gyri, the primary sensory-motor cortices, and the supplementary motor area) [19].

The literature on the role of RSNs in incidental recall paradigms is sparse. One study has investigated the role of RSNs in the incidental retrieval of episodic memories, concluding that relative deactivation of the DMN results in poor incidental recall scores at the time of retrieval [20]. These results confirm the function of the DMN, namely that suppressing it serves to reduce task-irrelevant processing during sensory intensive tasks. While brain activity during incidental recall appears to be TPN dominant (given the high attentional demand of such a task), little is known about how RSNs influence incidental recall at the time of encoding. According to Craik and Lockhart's levels of processing effect, accurate incidental recall of semantic memories is strongly influenced by modulatory variables at the time of encoding. For example, the familiarity of a semantic memory will influence the accuracy by which it can be successfully incidentally recalled [5]. Additionally, memories with greater mental imagery vividness also tend to be recalled at a higher percentage as imagery vividness is often tied to emotional salience [21]. Such modulatory variables are thought to influence the dominant RSN activation pattern at the time of memory encoding [22,23].

Despite the generalized anti-correlated nature of the DMN and the TPN, significant variation in task-dependent RSNs have been observed in both experimental conditions involving healthy controls [24,25] and in numerous psychological disorders [26–36]. Task-based RSN variation in healthy controls is thought to represent inherent genetic variability [37], with some individuals demonstrating dominant TPN activation, others demonstrating dominant DMN activation and some reporting mixed RSN activation for a given task.

One such task that elicits large individual variation in dominant RSN activation is visual mental imagery. Very recently, it has been shown that individuals capable of highly vivid imagery visualization capabilities overwhelmingly utilize the TPN, while individuals with low mean vividness ratings overwhelmingly utilize the DMN [38]. Similarly, in memory recognition paradigms the strength associated with stimulus familiarity demonstrates a dichotomous RSN activation pattern. Specifically, individuals with strong familiarity judgments at the time of stimulus presentation demonstrate a dominant TPN activation, and individuals with weak familiarity judgments demonstrate a dominant DMN activation [39]. Research on the relationship between the strength of neuropsychological variables (in this case familiarity and vividness), at the time of stimulus encoding, and their corresponding RSN activation is of critical importance to understand how learning occurs at the level of large-scale neural networks. Our study is among the first to investigate how the strength associated with traditional modulatory psychological variables affects RSN activation and whether differential RSN activation has significant effects on memory consolidation.

In the present study, we report a secondary data analysis on a corpus of cuing-nouns, which examines how the naturally varying strength of word familiarity and imagery vividness influences performance in an incidental free-recall task. We entertain a novel theory as to how VIR can be explained, incorporating aspects of both dual-rote associative models and neurobiologically-inspired NMPH. We propose that VIR is best explained by a dualistic neuropsychological model mimicking the global activity in two large RSNs, the DMN and the TPN. Our proposed model of memory consolidation posits that it is not solely the strength of neuropsychological variables that influences future memory consolidation, it is the dominant pattern of RSNs at the time of encoding that accounts for most of the variability in memory consolidation scores. We hypothesize that dominant RSN activity plays a critical role in memory consolidation given the substantial findings in the clinical literature which demonstrate that abnormal RSN connectivity, particularly high frequencies of DMN and TPN activation, result in memory deficits [24–34]. As a result of these findings, our model hypothesizes that conditions in which stimulus encoding reflects competing activation of DMN and TPN will result in poor memory consolidation scores. That is, two encodings, even if weak, will result in better learning outcomes compared to a mix of one strong and one weak encoding.

The approach of testing our dualistic RSN neuropsychological model of memory consolation by relying on psychological evidence is supported by previous research which has established that some psychological variables of low strength have the ability to selectively engage the DMN, while other psychological variables of high strength have the ability to selectively engage the TPN. Overall, our study aimed to investigate how RSN variability at the time of encoding affects memory consolidation, as measured by scores on an incidental recall task. We expected to find that higher RSN variability/competition, represented by noun-cues associated with mixed scores on our neuropsychological variables (low vividness and high familiarity, or vice versa), would be associated with poor memory recall, whereas congruent levels (high vividness and high familiarity; low vividness and low familiarity) would be associated with higher memory recall (Figure 1).

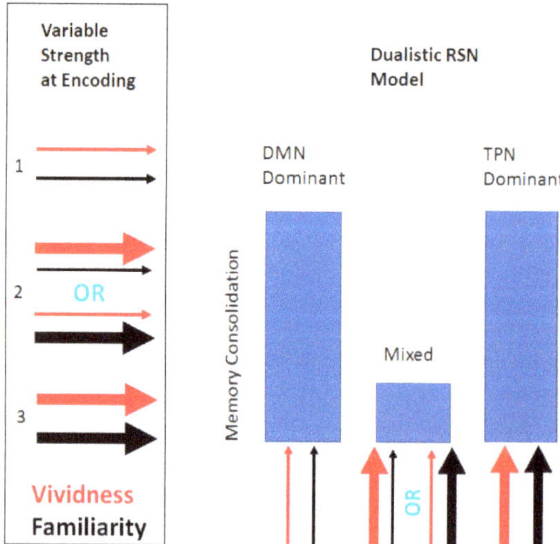

Figure 1. Schematic depicting the dualistic resting-state networks (RSN) neuropsychological model of memory consolidation. The model depicts its predicted memory consolidation score in relation to the three possible combinations of variable strength at the time of encoding. The width of the arrows indicates the strength of the variables at encoding.

The proposed analysis permits us to contrast our predictions using our dualistic RSN neuropsychological model with those derived from Paivio's dual coding theory and the NMPH, as they relate to VIR. Since as mentioned, NMPH proposes to explain (i.e., reduce) retrieval in terms of basic neurobiological mechanisms, such a framework, can be extended by incorporating current brain imaging research which indicates that higher mental imagery vividness ratings are associated with an overall increase in neural excitability in the relevant cortical areas [40]. Similarly, familiarity related cortical regions, such as the left angular gyrus, have been shown to increase in neural excitability when participants report that a memory is experienced as more familiar [41]. Thus, relating the above neural excitability patterns with the observed memory performance, if NMPH were to be correct, we would expect higher recall probabilities from noun-cues associated with high vividness and high familiarity ratings (high neural excitability). We would expect low recall probabilities from noun-cues associated with low vividness and high familiarity ratings or vice versa (moderate excitability), and we would expect average recall probabilities from noun-cues associated with low familiarity and low vividness ratings (low neural excitability). In contrast, if Pavio's dual coding theory were to be correct, we would expect that a higher probability of recall would manifest when both types of memory representations, imagery vividness and imagery familiarity, are strengthened. The recall probability should increase monotonically in relation to the strength of both imagery vividness and imagery familiarity (for comparison of model predictions see Figure A1).

2. Method

2.1. Design and Analytic Strategy

The present investigation consisted of secondary analyses of a corpus of well-characterized stimuli. The analyses involved a by-item approach followed by a confirmatory linear mixed logistic regression model.

In the by-item analysis, we applied a mixed ANOVA model on the means of recall probability and incidental noun recall and image latency times (RTs) for the stimulus words collapsed across all participants. This procedure, which is ordinarily the most widely used for by-item analysis [42], permits to avoid violating the assumption of independence needed to perform statistical hypothesis testing. By averaging all observations for each stimulus word, it was ensured that only one instance of a participants' data was used per stimulus word. Thus, the stimuli were the units of analysis, as they were treated as random variables (as if they were "subjects"). The generalizability of results, therefore, referred to both subjects and items populations, washing out individual difference effects (see [42]).

To confirm the ANOVA model, the linear mixed logistic regression approach consisted of analyzing each individual observation nested within participants and stimuli, instead of comparing the averaged responses by stimulus word. This supplementary method was adopted because of its ability to account for within-subjects effects, thereby enabling statistical testing within and between subjects without violating the assumption of independence, further allowing for stronger statistical power than the ANOVA model. (For details on the particular use of linear mixed regression models followed here see [43]).

The database used in this study is available on the archived website [44].

2.2. Stimuli

2.2.1. Initial Stimulus Selection

Fifty noun cues were selected from an initial body of 150 noun description-cues from previous extensive research [3,45–48]. The selection of the nouns for inclusion in the present database was operated through a series of stages. The first stage was one of data reduction to minimize the possibility of confounding attributes. The stimuli in the initial set were saved with an id number and variable columns of attribute variables in an excel file and successively underwent a series of automatic match via the merge command using IBM SPSS Statistics version 25 (Chicago, IL, USA) with regards to noun or compound word frequency, imageability, concreteness, emotional valence (all neutral), and readability attributes all drawn from updated online version of the dictionary file from the Medical Research Council (MRC) Psycholinguistic Database Version 2.0 (MRC2.DCT) [49].

The processed items included single and two-noun descriptions comprising both animate (e.g., dog, cat) and inanimate objects (e.g., car, bottle). To be included in the final database, the stimuli had to be above the mean imageability concreteness and meaningful attributes. All MCR values lie in the scaled range 100 (actual rating of 1) to 700 (actual rating of 7) with the maximum entry of 660 (i.e., 6.6), a mean of 490 (i.e., 4.9) and a standard deviation of 99 (i.e., 0.99). Extensive research in our lab revealed no reliable differences between these two subsets of stimuli in terms of the vividness or latency of elicited imagery. Other secondary analyses indicated that these descriptions are generally rated as relatively emotionally neutral, with negligible inter-item variability along a simple emotional rating scale [50]. These stimuli underwent further automatic merge and selection for several other aspects known to have potentially confounding correlations with other factors in the study. Verbal cues with higher concreteness levels are recalled at significantly higher rates [51,52]. Imageability, which refers to how easily a mental image can be generated from a word, correlates with concreteness [53]. We used the norms reported in the MRC2.CTC to confirm and validate that the selected cuing words had approximately the same scores on these factors [54]. This indirectly controlled for age of acquisition (average age a word enters a subject's lexicon), as the latter is very strongly predicted by both imageability and concreteness [55,56]. Nonetheless, we still used age of acquisition score norms from the MRC Psycholinguistic database to check for potential confounding effects. All diagnostic analyses showed age of acquisition had no significant or relevant effects. Hence, this variable was dropped from further analysis. At the end of this initial stage, only sixty descriptions were retained.

The second stage involved the collection of direct vividness ratings from raters as part of an imagery and incidental recall experiment using the sixty items which survived the first selection stage. It is important to point out that the MRC2.DCT does not contain vividness norms for the body of nouns contained in it. The procedure and protocol used in the stage are described in detail in the following section.

2.2.2. Vividness Rating Procedure

Since vividness was our main independent variable and was not derived from commonly available databases, we here report a summary of the rating procedure used to obtain the corresponding data for the sixty stimuli which survived the first selection stage, which is graphically represented in Figure 2A. This protocol was modeled after the paper and pencil procedures used in the normative studies merged in the MRC2.DCT (additional details of the experimental procedures can be found in [3]). The protocol was approved by the Carleton University Research Ethics Board, in strict adherence with the Tri-Council Policy Statement: Ethical Conduct for Research Involving Humans [57].

Sample of Vividness Raters

Participants serving as raters were 26 first-year university students (age range = 17–25; 14 female and 12 male). None had participated in an imagery study before. Participants signed up through a subject pool within 3 weeks of beginning introductory psychology courses, with 2% credit toward their final grade used as an incentive. No significance was found for gender or age against any factors, so these variables were dropped from further consideration. Participants under the age of 18 had to provide a written informed consent letter signed by their legal guardian to participate.

Image Generation Phase

Participants were seated facing a computer monitor and pressed the right mouse button to begin each trial. Upon clicking the mouse, an alerting beep was sounded, followed 250 ms later by the display of a noun-cue at the center of the screen. Participants were instructed to read the cue silently and as quickly as possible. They were immediately asked to generate an image that corresponded to the noun-cue. When participants felt that their mental image generation was at its most vivid state, they pressed the right mouse button. Upon pressing the button, another alerting beep was sounded, followed 250 ms later by a horizontal array of seven choices appearing near the bottom of the screen. From left to right, each button was labeled with one of seven vividness level descriptions in a seven-point scale format: ((1), "no image"; (2), "very vague/dim"; (3), "vague/dim"; (4), "not vivid"; (5), "moderately vivid"; (6), "very vivid"; and (7), "perfectly vivid"), as seen in previous research [47,58]. Participants were familiarized with the rating system during pre-test practice sessions. Participants used the mouse to click on one of these seven buttons and were instructed to rate any failure to generate an image as a "no image." There was no deadline for their response.

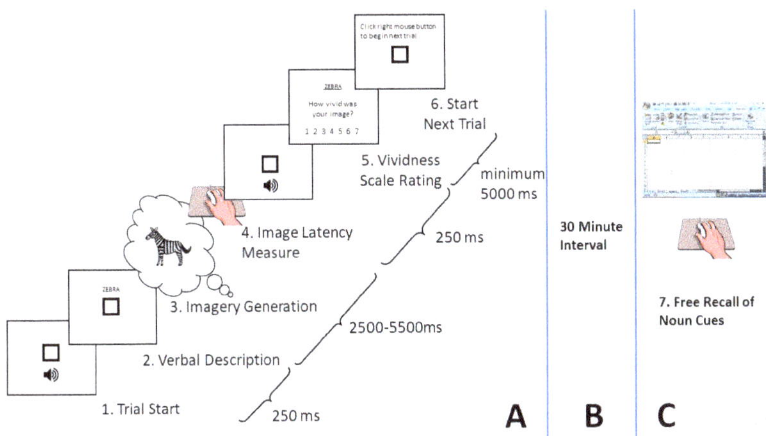

Figure 2. Schematic depicting the overall design of the imagery-generation incidental recall task. (**A**) Participants pressed the right mouse button to begin each trial (1). Upon clicking the mouse, an alerting beep was sounded, followed 250 ms later by the display of a noun-cue at the center of the screen (2). Participants were instructed to read the cue silently and as quickly as possible. They were immediately asked to generate an image that corresponded to the noun-cue (3). When participants felt that their mental image generation was at its most vivid state, they pressed the right mouse button (4). Upon pressing the button, another alerting beep was sounded, followed 250 ms later by a horizontal array of seven choices appearing near the bottom of the screen (5). From left to right, each button was labeled with one of seven vividness level descriptions in a seven-point scale format: ((1), "no image"; (2), "very vague/dim"; (3), "vague/dim"; (4), "not vivid"; (5), "moderately vivid"; (6), "very vivid"; and (7), "perfectly vivid"). Following the vividness response during the rating procedure, the array of buttons disappeared, and the display reverted back to a screen instructing the participant to click the mouse when they were ready to begin the next trial (6). A minimum of 5 s was needed between vividness response and the start of the next trial. (**B**) After completing the image generation phase, participants were told to take a break and fill out paperwork, including a debriefing session. (**C**) Exactly 30 min from their last trial, participants were asked to recall as many of the noun cues as possible on a blank excel spreadsheet (7).

Stimulus Familiarity Matching and Diagnostic Procedure

The third stage of the selection involved finding a complete archival match of familiarity for the sixty stimuli, again using the MRC2.DCT and using the same merging procedure outlined in Section 2.2.1. to consolidate the present database. MRC2.DCT is an online dictionary file being provided for public research use along with some programs which can be used either to access the dictionary or as examples on which to model programs which match users' specific needs. The dictionary file does not contain any information which is original to it but was assembled by merging a number of smaller databases published in the psycholinguistic and imagery literature [59]. The original procedure for rating the items consisted of paper and pencil protocol similar to the one we used, albeit in computerized form. In the original norms, the equivalent range of the ratings was 1.00 to 7.00. This database dictionary differs from other machine usable dictionaries in that it includes not only syntactic information but also psychological data for the entries. The file contains 9392 words which possess imagery and other attributes and familiarity ratings except for vividness. The columns 26 to 28 labeled as "FAM" Familiarity stands for 'printed familiarity'. The FAM values were derived from merging three sets of familiarity norms: Paivio, Yuille and Madigan, Toglia and Battig, and Gilhooly and Logie [60–62]. The method by which these three sets of norms were merged is described in detail in Appendix 2 of the MRC Psycholinguistic Database User Manual [63]. FAM values lie in the range 100 to 700 with the maximum entry of 657, a mean of 488, and a standard deviation of 99: Note that they are integer values.

The fourth and final stage involved analysis of the distribution of the vividness and familiarity values to avoid range restriction and diagnostics to eliminate outliers. The latter stage narrowed the final number of stimuli in the database further to fifty.

Filler Stimuli

In addition, we selected ten other noun-cues from previous earlier research (Paivio, Yuille, and Madigan in 1968) to use as buffer items during the incidental recall phase of the experiment (i.e., to filter out recency and primacy effects during recall) [60]. The sixty cues were presented in random order, preceded by 4 buffer noun-cues and followed by 4 other buffer noun-cues (which were presented in a fixed order).

2.3. Dependent Variables

2.3.1. Image Latency Times Measure

A main dependent variable in our study was image latency time. Following the vividness response during the rating procedure, the array of buttons disappeared, and the display reverted back to a screen instructing the participant to click the mouse when they were ready to begin the next trial. In an effort to minimize imagery persistence between trials, stimuli were presented in random order with a minimum inter-trial interval of 5 s, as was done in Craver-Lemley and Reeves [64]. Participants were not informed that latency times (RTs) were covertly measured from when the stimulus was presented to when they gave the first response prompting the appearance of the vividness buttons screen (see (4) in Figure 2). Button presses were justified as the way of communicating to the experimenter a complete image was formed, which was ready to be rated, and prompted the appearance of the vividness scale buttons.

2.3.2. Free Incidental Recall Measure

The main dependent measure in our study was correct recall rates of the noun cues presented during the vividness rating procedure. After completing the image generation phase, participants took a 20 min break. Afterward, they were asked to return to the lab to fill out additional paperwork, to receive course credit, and complete the debriefing process. Prior to the image generation phase, participants had not been informed that they would be required to recall any of the stimuli. Upon their return, precisely 30 min from the end of the image generation phase, they were asked to complete the incidental recall task, wherein they were required to recall and record as many of the previously read noun cues as possible on a blank excel spreadsheet.

Each phase of the previously described procedures was exclusively conducted by one of two paid undergraduate research assistants. Both research assistants received training in their module, yet were unaware of the specific purposes and hypotheses of the study.

3. Results

In this section, before reporting the results related to the main hypothesized effects, we describe the characteristics of the by-item analysis in terms of descriptive and correlational statistics.

3.1. Stimuli Characteristics: Descriptive and Correlations

For an illustration of the relationship between variables, tables are shown in the following section. Table 1 contains descriptive information of the variables under considerations, including scores of the variables measured. Table 2 shows the correlations between all variables.

Table 1. The descriptive information of the experimental variables.

Variables	M	SD	Min	Max
Vividness mean	5.26	0.52	3.99	6.12
RTs (ms)	3892.59	524.24	2692.87	5271.39
Familiarity mean	5.60	0.74	3.82	6.84
Recall probability	0.40	0.22	0.08	0.9231

Vividness and familiarity ratings: Min = 1 to Max = 7. RTs = Image latency times in milliseconds.

Table 2. Pearson Correlations between the experimental variables.

Variables	Vividness	Reaction Time	Familiarity	Recall Probability
Vividness		−0.39 **	0.48 **	−0.16
Reaction time	−0.39 **		0.04	−0.01
Familiarity	0.48 **	0.04		0.01
Recall Probability	−0.16	−0.01	0.01	

** $p < 0.001$ (2-tailed).

3.2. Mental Imagery Vividness and Familiarity

The mean vividness score across all noun-cues significantly correlated with each noun-cue's familiarity rating (r_p (50) = 0.48, p (two-tailed) < 0.001, $\eta_p^2 = 0.23$), as shown in Table 2. This correlation confirms the relationship between vividness and familiarity posited by previous research, suggesting that our dataset demonstrates the typical relationship observed between the two variables [7].

3.3. Mental Imagery Vividness and Reaction Time

The mean vividness score across all noun-cues significantly correlated with each noun cue's mean reaction time (r_p (50) = −0.39, p (two-tailed) < 0.001, $\eta_p^2 = 0.15$), as shown in Table 2. This correlation, known as the "vivid-is-fast" phenomenon, has been observed in previous research and therefore helps to validate our experimental dataset [65].

3.4. Effects of Resting State Networks on Learning Outcomes

3.4.1. By-Item Analysis Approach

Following our hypotheses, to assess how RSN activation pattern's influence mean recall probability across noun-cues, we devised a one-factor ANOVA with three levels, each representing distinct resting state activation patterns. As was described in the introduction, when a dominant DMN activation is observed in human participants, both mental imagery vividness and object familiarity are experienced consciously as a weak stimulus. Inversely, when a dominant TPN activation is observed, both mental imagery and object familiarity are experienced consciously as a strong stimulus. Given this relationship, it is possible to infer how RSNs affect learning outcomes by investigating the relationship between mean familiarity, mean vividness, and mean recall probability ratings. Applying a widely-used continuous variable partitioning technique [66], we created a variable with three levels that aimed to represent the following three distinct resting state network activation patterns: 1. TPN dominant activation (composed of noun cues which have high vividness and high familiarity ratings); 2. Mixed TPN and DMN activation (composed of noun cues with high vividness/low familiarity or low vividness and high familiarity); 3. DMN dominant activation (composed of noun cues with low vividness and low familiarity ratings). The decision to create a three-leveled variable characterized by the strength of mean familiarity and mean vividness scores was influenced by the NMPH model utilized in Norman and Newman [9]. In their model, the moderate/mixed level is of critical importance to assess non-monotonic relations between the strength of a stimulus and subsequent learning outcomes. For this reason, we judged that three levels would be ideal given that this number of levels would adequately contrast differences between states of stimulus competition (high/low) and states of stimulus collaboration. The descriptive information detailing the relationship between the resulting RSN variables and recall probability is shown in Table 3.

Table 3. Descriptive statistics of total noun-cues recalled (in percentages) as a factor of resting state network type.

			Descriptive		
Variables	N	**M**	**SE**	**Min**	**Max**
TPN dominant	18	0.40	0.06	0.08	0.92
Mixed	14	0.27	0.04	0.08	0.58
DMN dominant	18	0.49	0.04	0.15	0.77

The percentage of noun-cues recalled for each of the three RSN levels were entered on a 1 (recall probability) × 3 (RSN levels: low, mixed and high activation) one-way ANOVA. This procedure yielded a significant effect of resting state network type on recall probability ($F(2,47) = 4.51$, $p = 0.016$). Additionally, the ANOVA resulted in a large effect size, $\eta_p^2 = 0.16$ indicating that 16% of the variance in recall probability is attributable to dominant RSN activation. To investigate whether the mixed resting state activation level impaired recall probability, as previously hypothesized (i.e., our switching hypothesis), a polynomial quadratic planned contrast was further applied to the data. The contrast demonstrated a significant quadratic trend ($F(1,47) = 7.18$, $p = 0.01$, $\eta_p^2 = 0.13$). This quadratic trend can be clearly observed as represented in the continuous connecting line in Figure 3. To investigate how well our results fit within the dualistic RSN neuropsychological model of memory consolidation, we conducted Dunnett's post-hoc testing using the DMN-Dominant level as our main comparison criteria. Dunnett's post hoc test revealed that there were no significant differences between the DMN-Dominant and TPN-Dominant levels (Dunnett's $t(47) = 1.45$, $p = 0.154$), however there was a significant difference between the DMN-Dominant and the Mixed levels (Dunnett's $t(47) = 3.01$, $p = 0.004$). These results confirm the hypothesized quadratic pattern of the dualistic RSN neuropsychological model of memory consolidation, as shown in the (Figure 3).

To investigate if the quadratic trend was present across different percentiles of the recall probability distribution, the data were converted into proportions as detailed in Table 4. Planned quadratic proportion contrasts were performed on each third of the recall probability distribution. For both the top third (i.e., high recall probability) and middle (i.e., medium recall probability) percentile there were significant positive quadratic trend ($z = 3.24$, $p < 0.001$, $\eta_p^2 = 0.21$; and $z = 2.42$, $p = 0.008$, $\eta_p^2 = 0.12$, respectively). Conversely, the low recall group showed a significant negative quadratic trend ($z = -4.3464$, $p < 0.001$, $\eta_p^2 = 0.38$). To adjust for multiple comparisons, a two-tailed Bonferroni correction was applied to each contrast using 95% confidence intervals. Given that the data being used are proportions, the confidence intervals were treated as the hypothesis test wherein if an interval crossed the 0 threshold it would signify that the null hypothesis failed to be rejected and therefore, the contrast would be interpreted as not significant. The following equation was used to test for multiple comparisons: $\sum p\lambda \pm z_{\frac{\alpha}{2g}} \sqrt{s_p^2 \lambda^2}$, for further clarification on planned contrast in proportions see Rosenthal and R Rosnow (1985) and Hays (1994) [67,68]. The results of the two-tailed Bonferroni correction are as follows: High recall probability (0.445 ± (−0.301)), Medium recall probability (0.6031 ± 0.2477), Low recall probability (−1.0476 ± 0.2394). Each contrast's 95% confidence interval did not cross the 0 threshold, indicating that the quadratic contrast amongst each level of the recall variable were significant at $p < 0.025$. The two significant positive quadratic trends in the top 2 percentiles of the recall probability distribution in addition to the significant negative quadratic trend in the bottom percentile demonstrate that noun cues associated with competing RSN activation patterns result in poor recall probability.

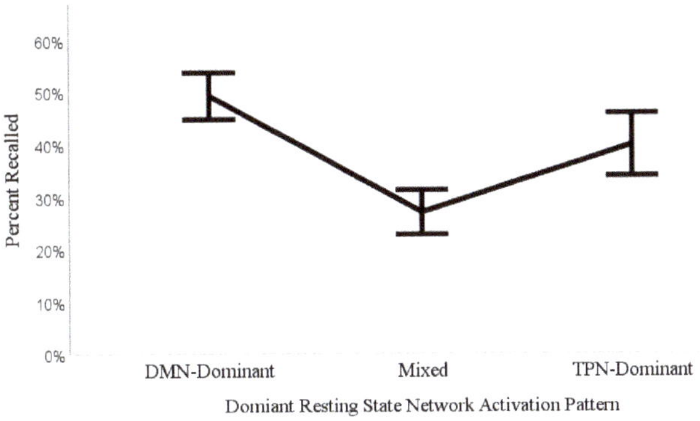

Figure 3. Percentage of words recalled in relation to the dominant resting state network activation pattern.

Table 4. Proportions and (total counts) of noun-cues recalled at three different types of resting state network action pattern.

Variables		DRSNAP			
	Variable Levels	TPN	Mixed	DMN	Total
Recall	High	0.06 (3)	0	0.1 (5)	0.16 (8)
	Medium	0.14 (7)	0.04 (2)	0.18 (9)	0.36 (18)
	Low	0.16 (8)	0.24 (12)	0.08 (4)	0.48 (24)
	Total	0.36 (18)	0.28 (14)	0.36 (18)	1.00 (50)

3.4.2. Linear Mixed Logistic Regression Approach

To assess how RSN activation patterns influence recall probability across each trial observation ($N = 1300$) we devised a linear mixed logistic regression model where recall of stimuli served as our binary target variable (1 = recalled, 0 = not recalled) and where dominant RSN activation patterns served as our fixed predictor variable. Similar to the by-item analysis, our RSN predictor variable contained three levels, each representing distinct resting state activation patterns. To demarcate each level of the predictor variable, we used the same partitioning technique used in the by-item analysis [66]. Specifically, the underling neuropsychological variable strength used to model the three RSN levels are as follow: 1. TPN dominant activation (composed of trial observations which have high vividness and high familiarity ratings); 2. Mixed TPN and DMN activation (composed of trial observations with high vividness/low familiarity or low vividness and high familiarity ratings); 3. DMN dominant activation (composed of trial observations with low vividness and low familiarity ratings). The descriptive information detailing the relationship between the resulting RSN variables and recall proportions are shown in Table 5.

Table 5. Descriptive statistics of total noun-cues recalled across all trials (in proportion) as a factor of resting state network type.

		Descriptive			
Variables	*N*	**M**	**SE**	**95% CI**	**95% CI**
TPN dominant	519	0.36	0.02	0.32	0.40
Mixed	528	0.29	0.02	0.25	0.33
DMN dominant	253	0.36	0.03	0.30	0.42

All values have been adjusted for within-subject effects.

Controlling for nested within-subjects data, we tested whether recalled noun-cues following a binomial logit link function could be predicted by the fixed effect categorical RSN predictor variable which reflected our three RSN levels (low, mixed, and high activation). This procedure yielded a significant fixed effect of resting state network type on noun-cue recall ($F(2,1297) = 3.30$, $p = 0.037$). To investigate whether these results supported the dualistic RSN neuropsychological model, we conducted Fisher's least significant difference (LSD) post-hoc testing. The LSD post-hoc test was used over other more conservative statistical tests for two primary reasons. First, since our predictor variable only consisted of three groups, the LSD procedure permits to preserve the experiment-wise type I error rate at nominal levels of significance, which nullifies the use of post-hoc tests that control for the family-wise error rate [69]. Second, for each level of our RSN predictor variable, the variance-covariance matrix demonstrated that covariance scores were all equal, and variance scores did not significantly differ among each other, suggesting that our data did not violate the assumption of compound symmetry needed for appropriate application of the LSD post-hoc test [70]. The LSD post-hoc test revealed that there were no significant differences between the DMN-Dominant and TPN-Dominant levels (LSD's $t(1297) = 0.178$, $p = 0.859$), however there were significant differences between both DMN-Dominant (LSD's $t(1297) = 1.97$, $p = 0.049$) and TPN-Dominant (LSD's $t(1297) = 2.31$, $p = 0.021$) when contrasted with the Mixed RSN level. These results confirm the hypothesized quadratic pattern of the dualistic RSN neuropsychological model of memory consolidation. Additionally, the results generated using the linear mixed logistic regression model helped to validate the findings derived from the by-item approach. Both statistical approaches demonstrated that learning outcomes are significantly predicted by the characteristic non-monotonic quadratic pattern of our dualistic RSN model of memory consolidation.

3.5. *Effects of Resting State Networks on Mental Imagery Latency (Reaction Time)*

The mean reaction times associated with each noun-cue for each of the three resting state network levels were entered in a 1 (reaction time) × 3 (resting state network levels: low, mixed, and high activation) one-way ANOVA. Results showed a significant effect of resting state network type on reaction time ($F(2,47) = 8.06$, $p = 0.001$). Additionally, the ANOVA resulted in a large effect size, $\eta_p^2 = 0.26$, indicating that 26% of the variance in reaction time is attributable to dominant resting state network activation. Dunnett's post hoc test revealed that visual mental image generation associated with the high strength level, reflecting TPN-Dominant activity, was significantly faster than visual mental image generation associated with both Mixed ($t(47) = 3.936$, $p = 0.0003$) and DMN-Dominant ($t(47) = 2.299$, $p = 0.013$) RSN activity levels (which did not differ between each other: $t < 1$). Additionally, we found equivalent results when testing whether imagery latency could be predicted by RSN activation, within and between each trial, by using a linear mixed regression model which assumed a normal distribution with an identity link (since this time our target was measured on a continuous scale). These results confirmed that image generation latency is quicker when the relevant imagery is associated with high strength.

4. Discussion

In the present study, we explored the degree to which imagery visualization and word familiarity affected learning outcomes in a traditional incidental recall paradigm. In doing so, the current study compared contrasting theoretical predictions regarding the relationship between the strength of neuropsychological variables and learning outcomes. Specifically, we contrasted associative theories that claim that the relative strength of neuropsychological variables should be additive [8], non-monotonic theories that claim that the relative strength of neuropsychological variables should be non-monotonic in relation to learning outcomes [9,10], and RSN theories which predict that learning outcomes should improve when competition between DMN and TPN is reduced [25,34,71].

Our main results demonstrate that recall probability is at its highest when the strength of our neuropsychological variables (noun familiarity/imagery vividness) are in a congruence at the time of stimulus encoding. In other words, when the strength of noun familiarity and imagery vividness are both strong or both weak, recall of the stimulus is at its highest probability. Using our dualistic RSN neuropsychological model, we were able to infer that the strength of our neuropsychological variables was associated with DMN activation when variables were of low strength and associated with TPN activation when variables were of high strength. When considering the inferred pattern of RSN activation, our results show that a competing activation pattern between DMN and TPN at the time of stimulus encoding results in significant impairment of recall probability.

An additional secondary finding was that the mean reaction time to complete the mental imagery task was quickest when the strength of our neuropsychological variables was high. In other words, it was quicker to generate visual mental images from noun cues under conditions where TPN was dominant at the time of encoding. The vivid-is-fast correlation between high image latency speed and high strength of neuropsychological variables has been reported in previous research [48,65,72,73]. However, our findings suggest that in addition to high strength variables, dominant-TPN at the time of encoding is also a key predictor of quick imagery generation latency. These results suggest an intriguing, novel explanation for the vivid-is-fast phenomenon. Namely, that it is not only the strength of neuropsychological variables that predict imagery generation latency but rather the dominant RSN pattern utilized during a mental imagery task. However, given the incongruent results between RSN patterns among our dependent variables (imagery latency was only significantly faster in the TPN grouping, while recall probability was significantly greater in both the DMN and TPN groupings), it is unclear whether image latency is related to the same processes responsible for memory consolidation. This is a research question open for future investigations.

In terms of the findings concerning the recall probability variable, our results do not support associative theories, as high memory consolidation is predicted to be exclusively a result of the additive strength of the preceding memory representations. Therefore, it can be concluded that Paivio's associative dual coding theory does not accurately model the incidental recall effect when operationalizing the strength of both mental imagery vividness and noun cue familiarity.

While our results did demonstrate a non-monotonic relationship between the strength of memory representations and recall probability, the directionality of our non-monotonic relationship did not align with that of Norman's non-monotonic-plasticity-hypothesis (NMPH). In the NMPH model, low strength memory representations result in lower memory consolidation scores compared to high strength memory representations. In our results, we found no significant differences between the high and low memory representations strength in terms of recall probabilities, suggesting that the NMPH may not accurately model memory consolidation at levels of reduction greater than single neurons. It is important to note that while our results did not precisely match the type of non-monotonic relationship that characterizes a neuron's level of excitation and the change in its associated synaptic strength, our results still demonstrated a non-monotonic relationship. This could suggest that there is, at the very least, a partial carryover effect of synaptic neuroplastic principles at the level of large scale distributed neural representations, given that non-monotonic results are rare within memory paradigms. However, the most likely interpretation of our results, that accounts for the largest amount of variability in our data, appears to be related to the variability in RSN dominance at the time of stimulus encoding. Utilizing the current RSN brain imaging literature [36,37] we were able to map the strength of our neuropsychological variables (imagery vividness/stimulus familiarity) with their associated dominant RSNs (DMN or TPN). This permitted the comparison of recall probability scores with the inferred dominant RSN at the time of encoding. Our data clearly suggest that when there is an antagonist relationship between the inferred DMN and TPN activation at the time of stimulus encoding, recall probability, and therefore, memory consolidation is significantly impaired.

Our overall findings help to confirm a hypothesis emerging from the mental health literature suggesting that a possible causal mechanism of memory deficits in neuropsychiatric disorders, such as depression, anxiety, attention-deficit-hyperactivity disorder, autistic spectrum disorder, and schizophrenia, are characterized by abnormalities in RSN activation [24–34]. For example, in schizophrenia, there is a significant increase in DMN activity during attention-demanding tasks, which normally recruits the TPN in healthy controls [74]. The increased DMN activation in schizophrenic patients results in a metabolic competition between the two anti-correlated RSNs, which is hypothesized to explain their stunted performance on such tasks [75]. The strongest evidence of the RSN memory interference hypothesis originates from research on ADHD. In a recent randomized, double-blinded, placebo-controlled clinical trial, researchers explored the RSNs of both ADHD patients and healthy controls while they performed attention-demanding tasks. Their results demonstrated that ADHD patients were unable to suppress the appropriate RSN both during the task and in-between trials. More importantly, ADHD patients were shown to have a significant amount of RSN activation variability during the task. This increased RSN variability, or increased competition between DMN and TPN, is hypothesized by the researchers to be responsible for the reduced scores on the behavioral task compared to controls [25]. Critically, however, these clinical findings provide general evidence that RSN homeostasis is important for efficient memory consolidation, but it is still unclear how abnormal RSN activation arises from such a wide variety of neuropsychiatric disorders. Hopefully, further experiments using healthy controls could help to determine the neurophenomenological marker's responsible for these core memory deficits.

To our knowledge, this study is the first to investigate how RSN variability might affect memory consolidation in healthy controls by manipulating the strength of neuropsychological variables that influence memory formation. While the results tend to validate the RSN memory deficit hypothesis in the mental health literature, there are still methodological limitations that should be considered prior to generalizing the results of this study. An initial limitation is a result of our reliance on indirect measurements of RSNs. While our dualistic RSN neuropsychological model is inferred from reliable brain imaging research [36,37], it is possible that the interaction between imagery vividness strength and stimulus familiarity at the time of stimulus encoding could reveal RSN activity which differs from our predicted inferences. Further studies using direct RSN imaging classification paradigms will need to be used to more precisely measure changes in dominant RSN activation in the function of both behavioral performance and of the strength of the chosen modulatory variables. A potential modulatory variable that could be used in future studies to help model the role of RSNs in memory consolidation is acute cannabis use. It has recently been demonstrated that cannabis is one of few pharmacological substances that has the effect of acutely increasing DMN connectivity during an attentionally demanding task [76]. Therefore, cannabis, in conjunction with psychological variables, such as mental imagery vividness and familiarly strength, could be used to artificially create an environment where individuals experience high degrees of competition between both the TPN and the DMN. Such experimental conditions could offer insight as to how various neuropsychological variables influence RSNs and therefore, memory consolidation processes.

5. Conclusions

In this study, we explored how varying degrees of both stimulus familiarity and mental imagery vividness ratings influenced verbal learning scores, as measured with an incidental recall task. Importantly, using relevant imaging research, we inferred dominant RSN activation patterns based on the reported strength of our neuropsychological variables at the time of stimulus presentation. We contrasted our results amongst three theoretical frameworks, namely: Paivio's associative theories, Normans' non-monotonic plasticity theory, and the newly postulated RSN memory interference hypothesis. The findings supported the latter hypothesis given that noun cues with the highest probability of being recalled were associated with noun cues which had the least amount of predicted RSN competition during stimulus presentation. The outcomes of this study, although preliminary, support the idea that a critical component of successful and efficient memory consolidation does not rely on the particular RSN present at the time of memory encoding (DMN or TPN), but rather it relies upon having minimal RSN activation variability between DMN and TPN at the time of memory encoding. This distinction could highlight relevant RSN biomarkers in the diagnosis of the mental health disorders affected by memory deficits. The RSN memory interference hypothesis could also help to illuminate our understanding of the broader neurophysiology of memory processes, complementing and expanding single-neuron neuroplasticity theories.

Author Contributions: Conceptualization, E.L. and A.D.; Data curation, E.L. and A.D.; Formal analysis, E.L.; Funding acquisition, A.D.; Investigation, E.L. and A.D.; Methodology, A.D.; Project administration, A.D.; Resources, E.L. and A.D.; Supervision, A.D.; Visualization, E.L.; Writing—original draft, E.L. and A.D.; Writing—review and editing, E.L. and A.D.

Funding: This research was partly funded by a grant from the Social Sciences and Humanities Research Council of Canada to A.D.

Acknowledgments: We thank Andrew Faulkner for handling preliminary set up of the corpus of data and database. We thank the two anonymous reviewers for their comments and statistical suggestions which have improved the paper.

Conflicts of Interest: The authors report no conflict of interest regarding the publication of this paper.

Appendix A

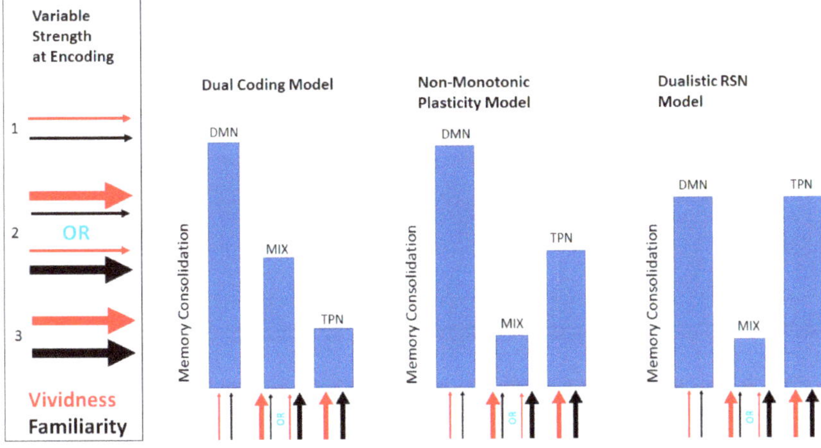

Figure A1. Schematic depicting the three contrasting models of memory consolidation. Each model depicts its predicted memory consolidation scores in relation to the three possible combinations of variable strength at the time of encoding. The width of the arrows signifies the strength of the variables at encoding.

References

1. Tulving, E.; McNulty, J.A.; Ozier, M. Vividness of words and learning in free-recall learning. *Can. J. Psychol. Can. Psychol.* **1965**, *19*, 242–252. [CrossRef]
2. Baddeley, A.D.; Andrade, J. Working memory and the vividness of imagery. *J. Exp. Psychol. Gen.* **2000**, *129*, 126–145. [CrossRef] [PubMed]
3. D'Angiulli, A.; Runge, M.; Faulkner, A.; Zakizadeh, J.; Chan, A.; Morcos, S. Vividness of Visual Imagery and Incidental Recall of Verbal Cues, When Phenomenological Availability Reflects Long-Term Memory Accessibility. *Front. Psychol.* **2013**, *4*, 4. [CrossRef] [PubMed]
4. Sheehan, P.W. The Role of Imagery in Incidental Learning. *Br. J. Psychol.* **1971**, *62*, 235–243. [CrossRef]
5. Yonelinas, A.P. The Nature of Recollection and Familiarity: A Review of 30 Years of Research. *J. Mem. Lang.* **2002**, *46*, 441–517. [CrossRef]
6. Mandler, G. Familiarity Breeds Attempts: A Critical Review of Dual-Process Theories of Recognition. *Perspect. Psychol. Sci.* **2008**, *3*, 390–399. [CrossRef] [PubMed]
7. Bellezza, F.S. Factors that affect vividness ratings. *J. Ment. Imag.* **1995**, *19*, 123–129.
8. Paivio, A. *Mental Representations: A Dual Coding Approach*; Oxford University Press: New York, NY, USA, 1990; Volume 10.
9. Newman, E.L.; Norman, K.A. Moderate Excitation Leads to Weakening of Perceptual Representations. *Cereb. Cortex* **2010**, *20*, 2760–2770. [CrossRef]
10. Lewis-Peacock, J.A.; Norman, K.A. Competition between items in working memory leads to forgetting. *Nat. Commun.* **2014**, *5*, 5768. [CrossRef]
11. Huijbers, W.; Schultz, A.P.; Vannini, P.; McLaren, D.G.; Wigman, S.E.; Ward, A.M.; Hedden, T.; Sperling, R.A. The encoding/retrieval flip: Interactions between memory performance and memory stage and relationship to intrinsic cortical networks. *J. Cogn. Neurosci.* **2013**, *25*, 1163–1179. [CrossRef]
12. Kim, H. Encoding and retrieval along the long axis of the hippocampus and their relationships with dorsal attention and default mode networks: The HERNET model. *Hippocampus* **2015**, *25*, 500–510. [CrossRef] [PubMed]
13. Block, N. Two neural correlates of consciousness. *Trends Cogn. Sci.* **2005**, *9*, 46–52. [CrossRef]
14. Fox, M.D.; Greicius, M. Clinical Applications of Resting State Functional Connectivity. *Front. Syst. Neurosci.* **2010**, *4*, 4. [CrossRef] [PubMed]
15. Greicius, M. Resting-state functional connectivity in neuropsychiatric disorders. *Curr. Opin. Neurol.* **2008**, *24*, 424–430. [CrossRef]
16. Fox, M.D.; Snyder, A.Z.; Vincent, J.L.; Corbetta, M.; Van Essen, D.C.; Raichle, M.E. The human brain is intrinsically organized into dynamic, anticorrelated functional networks. *Proc. Natl. Acad. Sci. USA* **2005**, *102*, 9673–9678. [CrossRef] [PubMed]
17. Smallwood, J.; Spreng, R.N.; Andrews-Hanna, J.R.; Andrews-Hanna, J.R. The default network and self-generated thought: Component processes, dynamic control, and clinical relevance. *Ann. N. Y. Acad. Sci.* **2014**, *1316*, 29–52.
18. Raichle, M.E. The Brain's Default Mode Network. *Ann. Rev. Neurosci.* **2015**, *38*, 433–447. [CrossRef]
19. Mantini, D.; Perrucci, M.G.; Del Gratta, C.; Romani, G.L.; Corbetta, M. Electrophysiological signatures of resting state networks in the human brain. *Proc. Natl. Acad. Sci. USA* **2007**, *104*, 13170–13175. [CrossRef]
20. Kompus, K. Default mode network gates the retrieval of task-irrelevant incidental memories. *Neurosci. Lett.* **2011**, *487*, 318–321. [CrossRef]
21. Dolcos, F.; LaBar, K.S.; Cabeza, R. Remembering one year later: Role of the amygdala and the medial temporal lobe memory system in retrieving emotional memories. *Proc. Natl. Acad. Sci. USA* **2005**, *102*, 2626–2631. [CrossRef]
22. Diana, R.A.; Yonelinas, A.P.; Ranganath, C. Imaging recollection and familiarity in the medial temporal lobe: A three-component model. *Trends Cogn. Sci.* **2007**, *11*, 379–386. [CrossRef] [PubMed]
23. St-Laurent, M.; Abdi, H.; Buchsbaum, B.R. Distributed Patterns of Reactivation Predict Vividness of Recollection. *J. Cogn. Neurosci.* **2015**, *7*, 1–19. [CrossRef] [PubMed]
24. Weissman, D.H.; Roberts, K.C.; Visscher, K.M.; Woldorff, M.G.; Woldorff, M. The neural bases of momentary lapses in attention. *Nat. Neurosci.* **2006**, *9*, 971–978. [CrossRef] [PubMed]

25. Mowinckel, A.M.; Alnæs, D.; Pedersen, M.L.; Ziegler, S.; Fredriksen, M.; Kaufmann, T.; Sonuga-Barke, E.; Endestad, T.; Westlye, L.T.; Biele, G. Increased default-mode variability is related to reduced task-performance and is evident in adults with ADHD. *NeuroImage Clin.* **2017**, *16*, 369–382. [CrossRef] [PubMed]
26. Uddin, L.Q.; Kelly, A.C.; Biswal, B.B.; Margulies, D.S.; Shehzad, Z.; Shaw, D.; Ghaffari, M.; Rotrosen, J.; Adler, L.A.; Castellanos, F.X.; et al. Network homogeneity reveals decreased integrity of default-mode network in ADHD. *J. Neurosci. Methods* **2008**, *169*, 249–254. [CrossRef] [PubMed]
27. Assaf, M.; Jagannathan, K.; Calhoun, V.D.; Miller, L.; Stevens, M.C.; Sahl, R.; O'Boyle, J.G.; Schultz, R.T.; Pearlson, G.D. Abnormal functional connectivity of default mode sub-networks in autism spectrum disorder patients. *NeuroImage* **2010**, *53*, 247–256. [CrossRef]
28. Sheline, Y.I.; Barch, D.M.; Price, J.L.; Rundle, M.M.; Vaishnavi, S.N.; Snyder, A.Z.; Mintun, M.A.; Wang, S.; Coalson, R.S.; Raichle, M.E. The default mode network and self-referential processes in depression. *Proc. Natl. Acad. Sci. USA* **2009**, *106*, 1942–1947. [CrossRef]
29. Zhao, X.-H.; Wang, P.-J.; Li, C.-B.; Hu, Z.-H.; Xi, Q.; Wu, W.-Y.; Tang, X.-W. Altered default mode network activity in patient with anxiety disorders: An fMRI study. *Eur. J. Radiol.* **2007**, *63*, 373–378. [CrossRef]
30. Liao, W.; Zhang, Z.; Pan, Z.; Mantini, D.; Ding, J.; Duan, X.; Luo, C.; Wang, Z.; Tan, Q.; Lu, G.; et al. Default mode network abnormalities in mesial temporal lobe epilepsy: A study combining fMRI and DTI. *Hum. Brain Mapp.* **2011**, *32*, 883–895. [CrossRef]
31. Bastos-Leite, A.J.; Ridgway, G.R.; Silveira, C.; Norton, A.; Reis, S.; Friston, K.J. Dysconnectivity within the default mode in first-episode schizophrenia: A stochastic dynamic causal modeling study with functional magnetic resonance imaging. *Schizophr. Bull.* **2015**, *41*, 144–153. [CrossRef]
32. Camchong, J.; MacDonald, A.W.; Bell, C.; Mueller, B.A.; Lim, K.O. Altered functional and anatomical connectivity in schizophrenia. *Schizophr. Bull.* **2011**, *37*, 640–650. [CrossRef] [PubMed]
33. Pomarol-Clotet, E.; Salvador, R.; Sarró, S.; Gomar, J.; Vila, F.; Martínez, A.; Guerrero, A.; Ortiz-Gil, J.; Sans-Sansa, B.; Capdevila, A.; et al. Failure to deactivate in the prefrontal cortex in schizophrenia: Dysfunction of the default mode network? *Psychol. Med.* **2008**, *38*, 1185–1193. [CrossRef] [PubMed]
34. Garrity, A.G.; Pearlson, G.D.; McKiernan, K.; Lloyd, D.; Kiehl, K.A.; Calhoun, V.D. Aberrant "Default Mode" Functional Connectivity in Schizophrenia. *Am. J. Psychiatry* **2007**, *164*, 450–457. [CrossRef] [PubMed]
35. Sheline, Y.I.; Raichle, M.E. Resting State Functional Connectivity in Preclinical Alzheimer's Disease: A Review. *Boil. Psychiatry* **2013**, *74*, 340–347. [CrossRef] [PubMed]
36. Sorg, C.; Riedl, V.; Mühlau, M.; Calhoun, V.D.; Eichele, T.; Läer, L.; Drzezga, A.; Förstl, H.; Kurz, A.; Zimmer, C.; et al. Selective changes of resting-state networks in individuals at risk for Alzheimer's disease. *Proc. Natl. Acad. Sci. USA* **2007**, *104*, 18760–18765. [CrossRef] [PubMed]
37. Glahn, D.C.; Winkler, A.M.; Kochunov, P.; Almasy, L.; Duggirala, R.; Carless, M.A.; Curran, J.C.; Olvera, R.L.; Laird, A.R.; Smith, S.M.; et al. Genetic control over the resting brain. *Proc. Natl. Acad. Sci. USA* **2010**, *107*, 1223–1228. [CrossRef] [PubMed]
38. Fulford, J.; Milton, F.; Salas, D.; Smith, A.; Simler, A.; Winlove, C.; Zeman, A. The neural correlates of visual imagery vividness—An fMRI study and literature review. *Cortex* **2018**, *105*, 26–40. [CrossRef] [PubMed]
39. Kim, H. Dissociating the roles of the default-mode, dorsal, and ventral networks in episodic memory retrieval. *NeuroImage* **2010**, *50*, 1648–1657. [CrossRef] [PubMed]
40. Belardinelli, M.O.; Palmiero, M.; Sestieri, C.; Nardo, D.; Di Matteo, R.; Londei, A.; D'Ausilio, A.; Ferretti, A.; Del Gratta, C.; Romani, G. An fMRI investigation on image generation in different sensory modalities: The influence of vividness. *Acta Psychol.* **2009**, *132*, 190–200. [CrossRef] [PubMed]
41. Kafkas, A.; Montaldi, D. Two separate, but interacting, neural systems for familiarity and novelty detection: A dual-route mechanism. *Hippocampus* **2014**, *24*, 516–527. [CrossRef]
42. Bedny, M.; Aguirre, G.K.; Thompson-Schill, S.L. Item analysis in functional magnetic resonance imaging. *NeuroImage* **2007**, *35*, 1093–1102. [CrossRef] [PubMed]
43. Hilbert, S.; Stadler, M.; Lindl, A.; Naumann, F.; Bühner, M. Analyzing longitudinal intervention studies with linear mixed models. *TPM Test. Psychometr. Method. Appl. Psychol.* **2019**, *26*, 101–119.
44. Mental Imagery Data from Words. Available online: https://zenodo.org/record/2658726#.XPptDsTQiUl (accessed on 5 February 2019).
45. D'Angiulli, A. Is the Spotlight an Obsolete Metaphor of "Seeing with the Mind's Eye"? A Constructive Naturalistic Approach to the Inspection of Visual Mental Images. *Imagin. Cogn. Personal.* **2009**, *28*, 117–135. [CrossRef]

46. D'Angiulli, A.; Reeves, A. Probing Vividness Increment through Imagery Interruption. *Imagin. Cogn. Personal.* **2003**, *23*, 63–78. [CrossRef]
47. D'Angiulli, A. Mental image generation and the contrast sensitivity function. *Cognition* **2002**, *85*, B11–B19. [CrossRef]
48. D'Angiulli, A.; Reeves, A. The Relationship between Self-Reported Vividness and Latency during Mental Size Scaling of Everyday Items: Phenomenological Evidence of Different Types of Imagery. *Am. J. Psychol.* **2007**, *120*, 521. [CrossRef]
49. MRC Psycholinguistic Database. Available online: http://websites.psychology.uwa.edu.au/school/MRCDatabase/uwa_mrc.htm (accessed on 13 April 2013).
50. D'Angiulli, A. Dissociating Vividness and Imageability. *Imagin. Cogn. Personal.* **2003**, *23*, 79–88. [CrossRef]
51. Paivio, A. *Imagery and Verbal Processes*; Holt, Rinehart & Winston: Oxford, UK, 1971.
52. Miller, L.M.; Roodenrys, S. The interaction of word frequency and concreteness in immediate serial recall. *Mem. Cogn.* **2009**, *37*, 850–865. [CrossRef]
53. Tse, C.-S.; Altarriba, J. Testing the associative-link hypothesis in immediate serial recall: Evidence from word frequency and word imageability effects. *Memory* **2007**, *15*, 675–690. [CrossRef]
54. Clark, C. *MRC Psycholinguistic Database*; The University of Western Australia School of Psychology: Perth, Australia, 1997.
55. Ma, W.; Golinkoff, R.M.; Hirsh-Pasek, K.; McDonough, C.; Tardif, T. Imageability predicts the age of acquisition of verbs in Chinese children. *J. Child Lang.* **2009**, *36*, 405–423. [CrossRef]
56. Barry, C.; Gerhand, S. Both concreteness and age-of-acquisition affect reading accuracy but only concreteness affects comprehension in a deep dyslexic patient. *Brain Lang.* **2003**, *84*, 84–104. [CrossRef]
57. Ethical Conduct for Research Involving Humans. Available online: http://www.pre.ethics.gc.ca/pdf/eng/tcps2-2014/TCPS_2_FINAL_Web.pdf (accessed on 15 Decemeber 2014).
58. Marks, D.F. New directions for mental imagery research. *J. Ment. Imag.* **1995**, *19*, 153–167.
59. Wilson, M. MRC psycholinguistic database: Machine-usable dictionary, version 2.00. *Behav. Res. Methods Instrum. Comput.* **1988**, *20*, 6–10. [CrossRef]
60. Paivio, A.; Yuille, J.C.; Madigan, S.A. Concreteness, imagery, and meaningfulness values for 925 nouns. *J. Exp. Psychol.* **1968**, *76*, 1–25. [CrossRef]
61. Toglia, M.P.; Battig, W.F. *Handbook of Semantic Word Norms*; Hillsdale, N.J., Ed.; Lawrence Erlbaum Associates: New York, NY, USA, 1978; ISBN 978-0-470-26378-5.
62. Gilhooly, K.J.; Logie, R.H.; Logie, R. Age-of-acquisition, imagery, concreteness, familiarity, and ambiguity measures for 1944 words. *Behav. Res. Methods* **1980**, *12*, 395–427. [CrossRef]
63. Coltheart, M. The MRC Psycholinguistic Database. *Quart. J. Exper. Psychol. Sect. A* **1981**, *33*, 497–505. [CrossRef]
64. Craver-Lemley, C.; Reeves, A. Visual Imagery Selectively Reduces Vernier Acuity. *Perception* **1987**, *16*, 599–614. [CrossRef]
65. D'Angiulli, A.; Reeves, A. Generating visual mental images: Latency and vividness are inversely related. *Mem. Cogn.* **2002**, *30*, 1179–1188. [CrossRef]
66. Preacher, K.J.; Rucker, D.D.; Maccallum, R.C.; Nicewander, W.A. Use of the Extreme Groups Approach: A Critical Reexamination and New Recommendations. *Psychol. Methods* **2005**, *10*, 178–192. [CrossRef]
67. Rosenthal, R.; Robert, R.; Rosnow, R.L. *Contrast Analysis: Focused Comparisons in the Analysis of Variance*; Cambridge University Press: Cambridge, UK, 1985.
68. Hays, W. *Statistics*, 5th ed.; Wadsworth Publishing: Fort Worth, TX, USA, 1994; ISBN 978-0-03-074467-9.
69. Meier, U. A note on the power of Fisher's least significant difference procedure. *Pharm. Stat.* **2006**, *5*, 253–263. [CrossRef]
70. Magezi, D.A. Linear mixed-effects models for within-participant psychology experiments: An introductory tutorial and free, graphical user interface (LMMgui). *Front. Psychol.* **2015**, *6*, 2. [CrossRef] [PubMed]
71. Xin, F.; Lei, X. Competition between frontoparietal control and default networks supports social working memory and empathy. *Soc. Cogn. Affect. Neurosci.* **2015**, *10*, 1144–1152. [CrossRef] [PubMed]
72. Rehm, L.P. Relationships among measures of visual imagery. *Behav. Res. Ther.* **1973**, *11*, 265–270. [CrossRef]
73. Hale, S.M.; Simpson, H.M. Effects of eye movements on the rate of discovery and the vividness of visual images. *Percept. Psychophys.* **1971**, *9*, 242–246. [CrossRef]

74. Van Buuren, M.; Vink, M.; Kahn, R.S. Default-mode network dysfunction and self-referential processing in healthy siblings of schizophrenia patients. *Schizophr. Res.* **2012**, *142*, 237–243. [CrossRef] [PubMed]
75. Whitfield-Gabrieli, S.; Ford, J.M. Default Mode Network Activity and Connectivity in Psychopathology. *Annu. Rev. Clin. Psychol.* **2012**, *8*, 49–76. [CrossRef] [PubMed]
76. Bossong, M.G.; Jansma, J.M.; van Hell, H.H.; Jager, G.; Kahn, R.S.; Ramsey, N.F. Default Mode Network in the Effects of Δ9-Tetrahydrocannabinol (THC) on Human Executive Function. *PLoS ONE* **2013**, *8*, e70074. [CrossRef]

Review

I Am Conscious, Therefore, I Am: Imagery, Affect, Action, and a General Theory of Behavior

David F. Marks

Independent Researcher, Arles, Bouches-du-Rhône, 13200 Provence-Alpes-Côte d'Azur, France; editorjhp@gmail.com

Received: 16 April 2019; Accepted: 8 May 2019; Published: 10 May 2019

Abstract: Organisms are adapted to each other and the environment because there is an inbuilt striving toward security, stability, and equilibrium. A General Theory of Behavior connects imagery, affect, and action with the central executive system we call consciousness, a direct emergent property of cerebral activity. The General Theory is founded on the assumption that the primary motivation of all of consciousness and intentional behavior is psychological homeostasis. Psychological homeostasis is as important to the organization of mind and behavior as physiological homeostasis is to the organization of bodily systems. Consciousness processes quasi-perceptual images independently of the input to the retina and sensorium. Consciousness is the "I am" control center for integration and regulation of (my) thoughts, (my) feelings, and (my) actions with (my) conscious mental imagery as foundation stones. The fundamental, universal conscious desire for psychological homeostasis benefits from the degree of vividness of inner imagery. Imagery vividness, a combination of clarity and liveliness, is beneficial to imagining, remembering, thinking, predicting, planning, and acting. Assessment of vividness using introspective report is validated by objective means such as functional magnetic resonance imaging (fMRI). A significant body of work shows that vividness of visual imagery is determined by the similarity of neural responses in imagery to those occurring in perception of actual objects and performance of activities. I am conscious; therefore, I am.

Keywords: vividness; mental imagery; consciousness; cognitive neuroscience; neuroimaging; cognitive psychology; behavior; verbal report; phenomenology; perception

---

1. Preliminaries

A General Theory of Behavior concerns the "I am" control center for (my) thoughts, (my) feelings, and (my) actions with (my) conscious mental imagery. Thus, 382 years after Descartes stated his "cogito, ergo sum" discovery, the evidence from cognitive neuroscience suggests the possibility of a more satisfying claim: "I am conscious, therefore, I am". The certainty of human existence relies not on the ability to think, but on the richer endowment provided by consciousness. It is not too large a stretch to assert that consciousness brought humans their privilege of pole position in the food chain. Without consciousness, humans would not benefit from what, I claim, is the pre-eminent force of nature that drives evolution, homeostasis. Organisms are adapted to each other and the environment as conscious beings because they possess an inbuilt striving toward stability, security, and equilibrium. Arguably, human homeostatic activity in niche construction generates less variation in the source of selection than where there is no feedback from organisms' activities to the environment [1].

The General Theory connects imagery, affect, and action with consciousness and the primary motivation that is psychological homeostasis. It is suggested that homeostasis provides a unifying concept across biology and psychology. Psychological homeostasis is as important to the organization of mind and behavior as physiological homeostasis to the organization of bodily systems. According to the theory, psychological homeostasis is available to organisms with consciousness, a process that embraces a near infinite supply of quasi-perceptual-affective images independently of the retina and sensorium. Homeostatic striving for security, stability, and equilibrium is a precondition for well-being. One major form of behavioral homeostasis is niche construction which alters ecological processes, modifies natural selection, and contributes to inheritance through ecological legacies. To do full justice to this very broad topic within the confines of the special issue, this invited review article refers to previous publications as supporting material, with all of the detailed references, a more complete explanation of the ideas, and the ongoing state of the research. One major purpose of the review is to present a summary of the evidence that mental imagery plays an essential role in the control of behavior. Mental imagery is an essential component of a new General Theory of Behavior involving homeostasis. A second purpose of this review is to explore the executive function of consciousness in the organization and control of behavior.

Central to the General Theory is the construct of mental image vividness, that combination of clarity and liveliness, so beneficial to consciousness and all its works: imagining, remembering, thinking, feeling, predicting, planning, pretending, dreaming, and acting. Introspective reports and objective indicators suggest the existence of wide individual differences in mental imagery vividness. Only in rare cases do is there evidence of absolutely no vividness at all. Vividness was studied under controlled conditions with standardized questionnaires, including the Vividness of Visual Imagery Questionnaire (VVIQ) [2–4]. An alternative method for studying vividness in experimental settings is to ask participants to provide vividness ratings (VR) on a "trial-by-trial" basis corresponding to the subjective experience at each particular moment in time [5]. This procedure avoids the problem leveled at the VVIQ that people cannot evaluate their private imagery along a common vividness scale because they have no objective reference points. However, this objection is refuted by validating data from controlled experiments. Assessment of vividness using introspective report was validated in multiple studies like those reviewed below using objective measures such as functional magnetic resonance imaging (fMRI). Also, the two end-points of the vividness scale are universally anchored by the descriptor "perfectly clear and as vivid as normal vision" at one end and "no image at all, you only 'know' that you are thinking of an object" at the other end of the scale.

Philosophers and psychologists discussing the vividness construct and its measurement sometimes displayed the characteristics of a "streetlight effect", the observational bias that occurs when searching for something and looking only where it is easiest, where there is light [6]. Searching in familiar terrains of logic and theory of mind while ignoring less familiar terrains of neuroscience and psychology can produce a fragmentary review and false conclusions. These problems come to the fore in a recent paper on "imaginative vividness" by Kind [7], in which the author suggests that it would be "best to retire our reliance on this notion entirely". These dismissive conclusions, I suggest, are based on incomplete examination of the evidence and faulty analysis of vividness and its phenomenology. For pragmatic reasons, I confine the present discussion to visual imagery, although similar principles are thought to apply to images of all modalities.

The General Theory assumes there is universal and constant striving for stability and equilibrium, a process termed "psychological homeostasis" [1]. Neuroscientific theories automatically fall under suspicion of material reductionism. However, a reductive approach is not required or desired and is not the approach taken here. The idea that behavior is reducible to physico-chemical reactions or to mechanistic "cogs and wheels" is rejected. This point is stated in the following working principle:

Working Principle: ***The voluntary behavior of conscious organisms is guided by a universal striving for equilibrium, a striving with purpose, desire, and intentionality.***

Psychological homeostasis, as a process of consciousness, is intentional, purposeful, and driven by the desire for security, safety, and equilibrium. It is necessary to assume that the mind/body system as a whole can be studied using objective methods. For one hundred years, mental imagery, vividness, and consciousness remained under-investigated in psychology and neuroscience for being too "subjective". The so-called "subjective" processes of vividness, mental imagery, and consciousness are amenable to objective study with tools and methods designed specifically for the purpose. This article indicates how these once "tabooed" areas became part of the scientific mainstream. A primary function of consciousness is the mental rehearsal of adaptive, goal-directed action through the experimental manipulation of perceptual-motor imagery. Every cycle of mental activity includes a goal, the means to reach it, and the consequences for the organism and objective world. The control system has a "meta level", a "schema level", and an "automatized level" in a hierarchical relationship. The meta level, which is a major function within "consciousness", sets the goals for a project of activity. The schema level monitors and controls the flow of action, making moment-by-moment adjustments in light of the goal set for it by the meta level above. The automatized level carries out repetitive and routine, everyday tasks that require zero planning or attention from higher up. When one hammers a nail into a piece of wood, one checks if the nail straight, if it is going in properly, whether it hit an obstacle, and so on. In cooking a soup, one tastes and seasons it with salt and pepper, not too much and not too little. In driving on the highway, one keeps to one's lane, attends to the road, monitors the other vehicles, and keeps to the speed limit, constantly vigilant for signs and warning signals, adjusting, adapting, and correcting. Conscious goal-setting lends purpose to our every action, striving for equilibrium with desire and intentionality.

2. Vividness

What exactly do I mean by "vividness"? It is necessary to be clear to avoid misunderstanding. I need search no further than the definition provided in an earlier publication: "a combination of clarity and liveliness. The more vivid an image, therefore, the closer it approximates an actual percept" [4]. The two elements, clarity and liveliness, are equally important. From the get-go, it is noted that vividness is independent of detail: there can be a lot of detail or none. Clarity and liveliness are the defining criteria. A vivid mental image of a red flag blowing in the wind requires nothing more than color, movement, and clarity—the quality of being clear, distinct, and intentional—a red flag, not a red rag. Liveliness is vitality, being animated, and the ability to evoke feelings of, say, attraction or repulsion. A vivid image is alive with vitality, energy, ebullience, and the potential to evoke activity and feeling.

A "life-like" image allows a person to experience it in a very real way. "Evoke" means to cause someone to sense or feel something. I appreciate this fact from personal experience. When I am watching a film of people lighting a camp fire, or smoking a cigarette, I actually can "smell" the smoke at the camp fire or the stench of a cigarette. These olfactory imaginings occur as if I am actually carrying out the activity, albeit on a lessor scale. Note that it is an activity, not a picture. Evoking a vivid image produces a life-like activity of "seeing", "hearing", "tasting", "smelling", "touching" or "feeling" something; mental imagery is a sensory-affective process that resembles, but is not identical to, perception with action.

Imagery, observation, and activity are produced by similar neural processes within a single system of representation in the brain. The evidence shows that the neurophysiological mechanisms that are active during physical skill acquisition are active during imagery and observation of the same skill [8]. Visual ideas may be cashed out as actions or they may be entirely covert. There is a ceaseless progression of ideas and associations in consciousness with or without the path markers of vivid imagery. In truth, every person has an Angie Thomas or a Khaled Hosseini sitting inside. Without vividness, however, no *The Hate U Give* or *Kite Runner*. There would also be fewer scientific discoveries—no Maxwell's demon, Einstein's elevator, or Schrödinger's cat (Figure 1).

Figure 1. Leonardo da Vinci, Ramón y Cajal, Marie Curie, and Albert Einstein—creative people who used vivid mental imagery to make world-changing discoveries. Einstein's thought experiments and his statements on the imagination are particularly salient.

Whatever else humans can do, our visual imagery ability is vital to our humanity, whether it is the everyday problem-solving required for stability, safety, and survival at home, in the workplace, or in the community. The essentially visual nature of thinking is "reflected" in the words used in conversation about "seeing", "views", "standpoints", "outlooks", "perspectives", "prospects", "angles", and "horizons", to mention only a small sample. Antonio Damasio explains the value of mental imagery to "creative intelligence" in human evolution: "Creative intelligence was the means by which mental images and behaviors were intentionally combined to provide novel solutions for the problems that humans diagnosed and to construct new worlds for the opportunities humans envisioned" [9] (p. 71).

The analysis of vividness risks entering a cul-de-sac when entertaining unhelpful metaphors such as the Platonic "picture" theory or the "descriptive", propositional theory [10] are adduced as if our mental hardware is akin to the technology for editing digital photographs or code. Mental images are not internal "pictures" or "photographs"; they are not anything like them. The mistake of the picture theory created insuperable problems and confusion. This review offers a different perspective on the nature and function of mental imagery, beginning with the measurement of vividness. It was demonstrated that conscious imagery is not equally vivid in all people and the reasons for, and consequences of, this fact stimulated a great deal of research. In the next section, I review one of the instruments employed in this research.

3. Vividness of Visual Imagery Questionnaire

To date, around 2000 studies have used the VVIQ or Vividness of Movement Imagery Questionnaire (VMIQ) [11] as a measure of imagery vividness. In this article, I address the VVIQ and leave the VMIQ to another occasion. The VVIQ is a self-report measure of the clarity and liveliness of visual imagery and, in so doing, aims to evoke images that vary in vividness, ambiance, and feeling as well. The instructions state the following:

"Visual imagery refers to the ability to visualize, that is, the ability to form mental pictures, or to 'see in the mind's eye'. Marked individual differences are found in the strength and clarity of reported visual imagery and these differences are of considerable psychological interest.

The aim of this test is to determine the vividness of your visual imagery. The items of the test will possibly bring certain images to your mind. You are asked to rate the vividness of each image by reference to the five-point scale given below. For example, if your image is 'vague and dim', then give it a rating of 4. After each item, write the appropriate number in the box provided. The first box is for an image obtained with your eyes open and the second box is for an image obtained with your eyes closed. Before you turn to the items on the next page, familiarize yourself with the different categories on the rating scale. Throughout the test, refer to the rating scale when judging the vividness of each image. Try to do each item separately, independent of how you may have done other items.

Complete all items for images obtained with the eyes open and then return to the beginning of the questionnaire and rate the image obtained for each item with your eyes closed. Try and give your 'eyes closed' rating independently of the 'eyes open' rating. The two ratings for a given item may not in all cases be the same." [4].

The five-point rating scale of the VVIQ is presented in Table 1. Some researchers prefer to reverse the numerical scale to make 5 = perfectly clear and as vivid as normal vision, and 1 = no image at all, you only "know" that you are thinking of an object.

Table 1. The rating scale in the Vividness of Visual Imagery Questionnaire [4].

Rating	The Image Aroused by an Item Might Be
1	Perfectly clear and as vivid as normal vision
2	Clear and reasonably vivid
3	Moderately clear and vivid
4	Vague and dim
5	No image at all, you only "know" that you are thinking of an object

The 16 items are arranged in blocks of four, in which each has a theme and at least one item in each cluster describes a visual image that includes movement (Table 2). Each theme provides a narrative to guide a progression of mental imagery. It is noted that at least one item in each cluster describes an activity or movement, indexing liveliness. The aim of the VVIQ is to assess visual imagery vividness under conditions which allow a progressive development of scenes, situations, or events as naturally as possible. The items are intended to evoke sufficient interest, meaning, and affect conducive to image generation. Participants rate the vividness of their images separately with eyes open and eyes closed.

Table 2. Items in the Vividness of Visual Imagery Questionnaire [4] *.

Item	Theme	Description
	Relative or friend †	For items 1 to 4, think of some relative or friend whom you frequently see (but who is not with you at present) and consider carefully the picture that comes before your mind's eye.
1	Relative or friend	The exact contour of face, head, shoulders, and body.
2 *	Relative or friend	Characteristic poses of head, attitudes of body, etc.
3 *	Relative or friend	The precise carriage, length of step, etc. in walking.
4	Relative or friend	The different colors worn in some familiar clothes.
	Natural scene: Rising sun	Visualize a rising sun. Consider carefully the picture that comes before your mind's eye.
5 *	Natural scene: Rising sun	The sun is rising above the horizon into a hazy sky.
6 *	Natural scene: Rising sun	The sky clears and surrounds the sun with blueness.
7 *	Natural scene: Rising sun	Clouds. A storm blows up, with flashes of lightening.
8 *	Natural scene: Rising sun	A rainbow appears.

Table 2. *Cont.*

Item	Theme	Description
	Shop	Think of the front of a shop which you often go to. Consider the picture that comes before your mind's eye.
9	Shop	The overall appearance of the shop from the opposite side of the road.
10	Shop	A window display including colors, shape, and details of individual items for sale.
11	Shop	You are near the entrance. The color, shape, and details of the door.
12 *	Shop	You enter the shop and go to the counter. The counter assistant serves you. Money changes hands.
	Natural scene: Lake	Finally, think of a country scene which involves trees, mountains, and a lake. Consider the picture that comes before your mind's eye.
13	Natural scene: Lake	The contours of the landscape.
14	Natural scene: Lake	The color and shape of the trees.
15	Natural scene: Lake	The color and shape of the lake.
16 *	Natural scene: Lake	A strong wind blows on the tree and on the lake causing waves.

* Eight of 16 items indicate activity or movement (marked *). † The first four items are from Peter Sheehan's (1967) shortened form of the questionnaire designed by Betts (1909).

For a small minority of people, the capacity for visual imagery is unavailable. In the absence of mental imagery, consciousness consists of "unheard" words, "unheard" music, and "invisible" imagery. This minority needs to employ more generic, verbal methods to recall events, and to plan goals and future activity—compensatory strengths that remain under-investigated.

4. The Nature and Function of Imagery

A large body of psychological and neuroscientific research is consistent with the tenet that imagery is functionally equivalent to, but not identical to, perception. Research summarized in a later section indicates similar anatomical patterning of neural activity in the cerebral cortex. I am discussing a graded phenomenon in which "the ordinary course of thought involves an interaction at sensory input with the central processes ... " [12] (p. 476), which can range from the very vivid to the completely abstract. As in perception, potentiality for action and a corresponding degree of anticipatory feeling and volition are part and parcel of a vivid mental imagery experience. Mental images come laden with associations in the form of feelings, e.g., attraction, calm, tranquility, fear, anger, or anticipation, useful "reflections" on life events in psychotherapy through image evocation in sessions of imagery therapy [13]. It is these attributes of mental imagery that enable works of literature to move readers to experience narrative and characters as if they are "real". Yet, it is vividness itself rather than arousal that is most closely correlated with the aesthetic appreciation of poetry, such as haiku or sonnets [14].

Another method to explore a person's imagery experience is to provide simple color suggestions while the participant looks into the center of a circle drawn on paper—the Open Circle Test [15]. When a vivid imager is invited to describe the imagery evoked by a color name, then a sequence of lively images may be experienced including isomorphisms and visual metaphors. For example, in one 38-year-old woman, the suggestion "yellow" produced a sequence starting with a yellow canary and transitioning into a variety of increasingly complex, dynamic images (Figure 2).

Each image connected and overlapped with the previous one. The sequence could have been continued indefinitely but we stopped after four transitions. The dynamic images triggered by a simple color name show the defining feature of liveliness. This attribute is neglected in practically all analyses of imagery. Vividness, action, and affect are the foundation stones for the General Theory of Behavior and of consciousness itself.

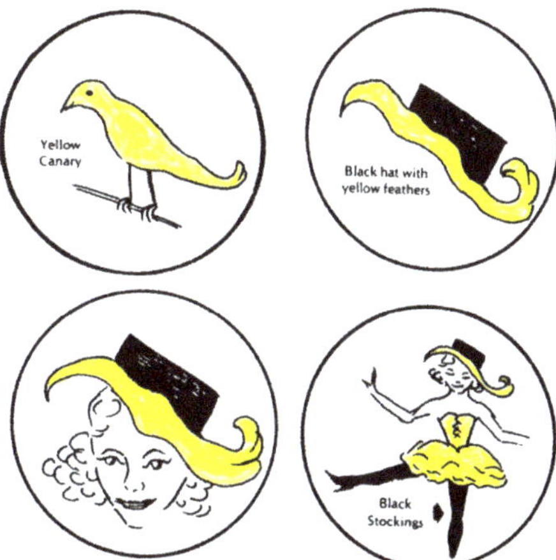

Figure 2. Sequence of projected visual images in response to the suggestion "yellow".

An extensive literature on mental "practice" refers to both imagery "rehearsal" and mental "simulation" [16,17]. Imagery is routinely and systematically employed in preparation and rehearsal of sports activity and was shown to produce enhanced performance across a wide variety of skillsets [18,19]. Studies of skilled performers show that activity cycles are more effectively rehearsed when they incorporate vivid imagery [20]. Studies of Olympic athletes and performers capable of specialist skills suggest that high imagery vividness is of most benefit to performances that have significant perceptual-motor components or require visualization of complex interactions at the object level [21].

Converging evidence suggests that mental simulation of movement and actual movement share similar neurocognitive and learning processes, leading to considerable interest in imagery simulation of movement as a therapeutic tool in the rehabilitation of stroke patients, patients with Parkinson's disease, and other neurological syndromes [22]. Conscious imagery enables the user to explore, select, and prepare physical and social activity. However, mental simulation is not a process of inspecting a holistic visual image like a picture or photograph in the "mind's eye". Mental simulations are constructed piecemeal, include non-visible properties, and can be used in conjunction with non-imagery processes, such as task decomposition and rule-based reasoning [23].

A common neural basis exists for imitation, observational learning, and motor imagery. During mental simulation, the excitatory motor output generated for executing the action is inhibited. The autonomic system is also activated during motor imagery. The principal function of consciousness is to plan and make predictions about the consequences of actions. Simulation enables the imager to mentally try out a sequence of goals, schemata, and actions that minimize hazard, loss, and pain.

The principal measures of vividness, the VVIQ and VR, are strongly associated with performance in different kinds of tasks: self-report, physiological motor, perceptual, cognitive, and memory [4,5,24,25]. To quote Runge et al. [5], "vividness can be considered a chief phenomenological feature of primary sensory consciousness, and it supports the idea that consciousness is a graded phenomenon". Recent research reviewed below showed that reported vividness is associated with early visual cortex activity relative to the whole-brain activity measured by functional magnetic resonance imaging (fMRI) and the performance on a novel psychophysical task.

Vividness of visual imagery correlates with fMRI activity in early visual cortex scores, demonstrating that higher visual cortex activity indexes more vivid imagery. Variations in imagery vividness depend on a large network of brain areas, including frontal, parietal, and visual areas. The more similar the neural response during imagery is to the neural response during perception, the more vivid or perception-like the imagery experience will be. From these findings, it can be concluded that an image is an idea with visual attributes. The more vivid the image, the more strongly a person will be aware of it. Upon reflection of the alternative actions available, it is possible to inhibit certain actions and implement others, or to keep actions "on hold" for the future. Thus, consciousness is able to facilitate successful striving toward goals, including the construction of new niches and environments, and thereby the effectiveness of type II homeostasis, providing a significant evolutionary advantage through niche construction and other adaptive mechanisms [1].

The evidence on the VVIQ [5,25] permits the following conclusions:

1. The VVIQ is only minimally contaminated or not contaminated at all by extraneous variables such as social desirability.
2. The VVIQ has acceptable levels of split-half and test-retest reliability.
3. The VVIQ has acceptable validity coefficients with other verbal-report measures of imagery and with physiological, motor, memory, and cognitive tasks.
4. The VVIQ has excellent criterion validity coefficients with perceptual tasks, exceeding those obtained with other self-report measures of imagery.
5. The ability to produce vivid visual imagery is of most benefit in tasks that are heavily loaded with perceptual-motor components and is of considerably less benefit to memory tasks.
6. A recent meta-analysis reported large effect size estimates (ESEs) for both VR and VVIQ measures, with larger ESEs for neuroscientific than behavioral-cognitive measures [5].

I summarize here two key findings.

STUDY 1: Vividness of mental imagery: individual variability can be measured objectively [26]. "When asked to imagine a visual scene, such as an ant crawling on a checkered table cloth toward a jar of jelly, individuals subjectively report different vividness in their mental visualization. We show that reported vividness can be correlated with two objective measures: the early visual cortex activity relative to the whole-brain activity measured by functional magnetic resonance imaging (fMRI) and the performance on a novel psychophysical task. These results show that individual differences in the vividness of mental imagery are quantifiable even in the absence of subjective report" [26] (p. 474).

"Results 3.1. Vividness of visual imagery correlates with fMRI activity in early visual cortex scores. We found a strong correlation (Figure 1C, $r = -0.73$, $p = 0.04$), demonstrating that higher relative visual cortex activity indexes more vivid imagery (a lower VVIQ score). This result suggests one can measure visual cortex activity to probe the vividness of a subject's imagery, thus obtaining a more objective measure of a previously subjective rating" [26] (p. 476).

STUDY 2: Vividness of visual imagery depends on the neural overlap with perception in visual areas [27]. I quote from the authors' abstract, as follows: "Research into the neural correlates of individual differences in imagery vividness point to an important role of the early visual cortex. However, there is also great fluctuation of vividness within individuals, such that only looking at differences between people necessarily obscures the picture. In this study, we show that variation in moment-to-moment experienced vividness of visual imagery, within human subjects, depends on the activity of a large network of brain areas, including frontal, parietal, and visual areas. Furthermore, using a novel multivariate analysis technique, we show that the neural overlap between imagery and perception in the entire visual system correlates with experienced imagery vividness. This shows that the neural basis of imagery vividness is much more complicated than studies of individual differences seemed to suggest" [27] (p. 1327).

"Significance statement: Visual imagery is the ability to visualize objects that are not in our direct line of sight—something that is important for memory, spatial reasoning, and many other tasks. It is known that the better people are at visual imagery, the better they can perform these tasks. However, the neural correlates of moment-to-moment variation in visual imagery remain unclear. In this study, we show that the more the neural response during imagery is similar to the neural response during perception, the more vivid or perception-like the imagery experience is" [27] (p. 1327).

"Results: To directly compare activity between perception and imagery, we contrasted the two conditions (see Figure 3). Even though both conditions activated the visual cortex with respect to baseline, we observed stronger activity during perception than imagery throughout the whole ventral visual stream. In contrast, imagery led to stronger activity in more anterior areas, including insula, left dorsal lateral prefrontal cortex, and medial frontal cortex . . . We modeled the imagery response for each vividness level separately." In Figure 4 the investigators plotted the difference between the main effect of perception and the main effect of imagery in the early visual cortex, for each vividness level.

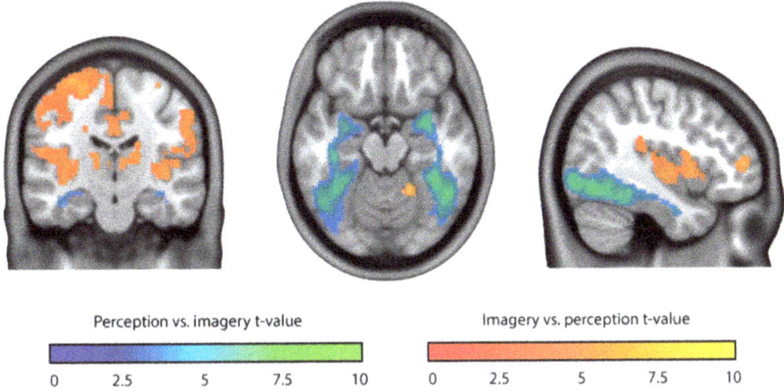

Figure 3. Perception versus imagery. Blue-green colors show t-values for perception versus imagery and red-yellow colors show t-values for imagery versus perception. Shown t-values were significant at the group level. Even though both conditions activated the visual cortex with respect to baseline, stronger activity occurred during perception than imagery throughout the whole ventral visual stream. In contrast, imagery led to stronger activity in more anterior areas, including insula, left dorsal lateral prefrontal cortex, and medial frontal cortex. Reproduced from Reference [27] with permission.

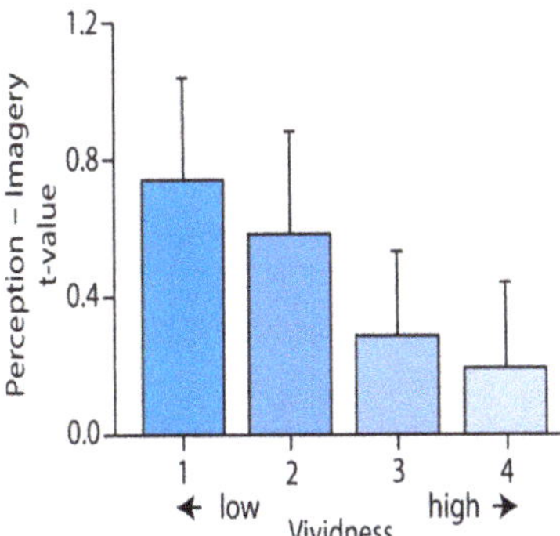

Figure 4. Difference between the effect of perception and the effect of imagery, separately for the four vividness levels. The higher the vividness is, the lower the difference between imagery and perception will be. In each trial, participants were shown two objects successively, followed by a cue indicating which of the two they subsequently should imagine. During imagery, a frame was presented within which subjects were asked to imagine the cued stimulus as vividly as possible. After this, they indicated their experienced vividness on a scale from one to four, where one was low vividness and four was high vividness. The results are shown for a voxel in the early visual cortex that showed the highest overlap between the main effect of perception and the main effect of imagery. More vivid imagery was associated with a smaller difference between perception and imagery [27] (p. 1335). Reproduced from Reference [27] with permission.

5. Action

"I always like to picture the game the night before: I'll ask the kitman what kit we're wearing, so I can visualize it. It's something I've always done, from when I was a young boy. It helps to train your mind to situations that might happen the following day. I think about it as I'm lying in bed. What will I do if the ball gets crossed in the box this way? What movement will I have to make to get on the end of it? Just different things that might make you one percent sharper," said Wayne Rooney [28].

The foundation for one's mental model of the world is coded in the central nervous system, which, among its many functions, represents the perceived and imagined worlds in various modalities of sensory-affective imagery. Visual imagery is understood to be a quasi-sensory experience which shares at least some of its generating processes with perception [29,30]. It is accepted that the perceptual system is inextricably linked to the action system, such that perceiving something often leads to some corresponding activity, either covert or overt [31]. Perceiving and imaging are not merely processes of identification brought about by looking and listening, but active performances in which specific intentions, purposes, and actions need to be fulfilled [32] (p. 6). In part, this is the distinction between "seeing what" and "seeing as". What something is seen as has implications for action. This fact is problematic for any theories in which perception and action are not functionally interlinked. If I am feeling thirsty in the desert and I see a reflection as a lake, I will move as rapidly as possible toward it feeling hopeful and encouraged. If I see the reflection as a mirage, I will try to ignore it and continue slowly on my way feeling discouraged. Note that the example includes an affective-motivational component, which is a characteristic of all types of activity and not just a selected example. We arrive at the activity cycle theory (ACT) [32,33] (Figure 5). The arrows in this and all the other figures in this article represent cause-and-effect connections, not correlational associations.

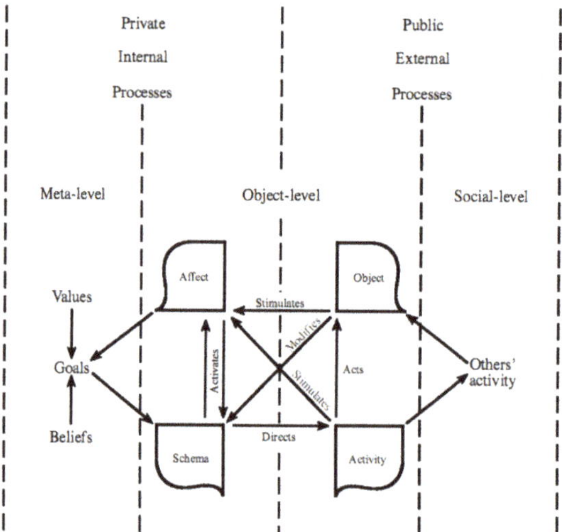

Figure 5. Activity cycle theory (ACT) consists of four systems, three representing a functional module for a psychological domain (affect, schema, activity) and one (object) representing physical objects or internal images and ideas. Private (internal) and public (external) processes are executed across the meta level, object level, and social level of organization. The system enables outcomes of alternative future actions to be appraised prior to committing to any course of action. Conscious imagery serves as a mental "toolbox", providing its internal contents for the user to explore in the selection and preparation of future physical and social activity. The affect system co-activates goals and schemata enabling the organism to strive toward desired outcomes. Please note that psychological homeostasis was not included in the ACT. However, the ACT formulation was a crucial step in formulating the General Theory of Behavior. Reproduced from Reference [33] with permission.

The activity cycle theory of conscious imagery claims that a primary function of consciousness is the mental rehearsal of adaptive, goal-directed action through the experimental manipulation of perceptual-motor imagery [33]. As predicted by this theory, meta-analyses [5,25] showed that the vividness of conscious mental imagery is strongly associated with precisely those performances most likely to benefit from the use of perceptual-motor imagery and mental practice. ACT helps to explain the existence and function of conscious experience.

The image is a cycle of activity triggered by any of four processes: object, affect, schema, and activity. As the evidence shows, the more vivid the image is, the more closely neural activation resembles that activated in real physical activity with actual objects. Covert actions are similar in neural terms to the state of execution of that action overtly. Mental imagery can be as simple as imagined looking, listening, or touching, or as complex as preparing a gourmet dinner or designing a scientific theory. The triggering schemata can be activated top-down or bottom-up. A cycle of mental activity always must include a goal, the means to reach it, and the consequences on the organism and objective world.

Evidence for the affective and somatic components of mental imagery is strong and was discussed for at least a century. In 1907, Wundt [34] wrote the following:

"When any physical process rises above the threshold of consciousness, it is the affective elements which, as soon as they are strong enough, first become noticeable. They begin to force themselves energetically into the fixation point of consciousness before anything is perceived of the ideational elements … They are sometimes states of pleasurable or unpleasurable character, sometimes they are predominantly states of strained expectation … Often there is vividly present … the special affective tone of the forgotten idea, although the idea itself still remains in the background of consciousness … In a similar manner … the clear apperception of ideas in acts of cognition and recognition is always preceded by feelings" [34] (pp. 243–244).

Silvan Tomkins taught us that the primary motivational system is the affective system, and that biological drives have impact only when amplified by the affective system [35]. A similar view was reached by Robert Zajonc [36]. When subjects imagine happy, sad, and angry situations, different patterns of facial muscle activity are produced that can be measured by electromyography [37]. People see objects with feeling. Affective representations of visual sensations are included in the brain's predictions of what sensations stand for and how to act on them in the future [38]. Similar affective responses occur when people mentally image not only faces, but more complex objects, scenes, and activities; however, the physiological responses are generally less intense in mental images [13].

6. Volition

Volition or will is the process by which an individual needs, wants, and commits to a course of action or goal. Volition is purposive striving, a primary psychological function for stability, security, and survival. Unsurprisingly, goal-directedness is at the root of many psychological theories and a significant feature of theories of consciousness [39,40]. In *The Theatre of Consciousness*, Bernard Baars [39] stated the following:

The only conscious components of action are as follows:

a. *the "idea" or goal (really just an image or idea of the outcome of the action);*
b. *perhaps some competing goal;*
c. *the "fiat" (the "go signal", which might simply be release of the inhibitory resistance to the goal);*
d. *sensory feedback from the action.*

Most actions are automatically generated. Only when there is novelty or special care required are actions conscious and under voluntary control. In his ideomotor theory of voluntary control, William James proposed that ideas and images trigger automaticity in brain centers that carry out voluntary actions. Conscious events are only needed to specify new goals and actions. When people focus on their movements, they may actually interfere with the automatic control processes that normally regulate movements, whereas focusing on the effect of movement allows the motor system to self-organize more naturally [41].

Affect, feelings, and drives are intimately related to the sense of striving to satisfy desires and goal attainment. Volition is the driving force of mental image generation and the contents of consciousness [42]. William James stated that "the pursuance of future ends (goals) and the choice of means for their attainment [schema] are the mark and criterion of the presence of mentality (consciousness) in a phenomenon" [42] (p. 8). Goals are set at a meta level governed by drives, values, and beliefs to allow planning of actions, inhibition of actions, and reflection ("wait and see") as the situation requires. A similar theory was independently developed by Marc Jeannerod (1999) [43].

The evidence suggests that consciousness evolved earlier in human evolution than language [44]. How much earlier is a matter for debate. Feinberg and Mallatt [45] argued that consciousness evolved 520 to 560 million years ago. Developmentally, also, consciousness is accessible earlier than language [44]. For these reasons, language and speech are subservient to imagery. Among its many uses, vivid imagery plays a key role in planning goal-directed behavior. The cognitive system has a meta level to control and monitor the object level. This duality of levels is advantageous because it enables moment-by-moment adjustments to goal-seeking behavior at the object level. Mental imagery instantiates the conscious representation of the self—the "I" of "I am"—and the environment, and interactions between self and others. Conscious imagery allows the conduct of mental simulations of action sequences at the object level without energy expenditure or risk. The object level interfaces with the social level in the public domain of shared activities and object levels. The possible outcomes of alternative future actions may be appraised prior to a course of action. In this way, conscious mental imagery serves as a mental toolbox, producing its internal contents for the user to explore and manipulate in the selection and preparation of future physical and social activity. This idea was anticipated by Francis Bacon more than 400 years ago when he wrote the following:

"For sense sendeth over to imagination before reason have judged; and reason sendeth over to the imagination before the decree can be acted; for imagination ever precedeth voluntary action" [46].

The principal role of mental imagery is to perform "thought experiments" by rehearsing activation of schemata and simulating alternative cycles of action to evaluate potential outcomes in the objective world before making actions physically. Conscious imagery provides the necessary competence to perform thought experiments in a risk-free manner. Thought experiments enable the imager to generate a sequence of interacting processes consisting of goals, schemata, actions, objects, and affects. The envisaged processes are similar to those of the scientist establishing theories and hypotheses and testing them using controlled investigation. The aims, methods, and results of the experiment are at the meta level and are determined by needs, beliefs, and values. The remaining processes are at the object level. Once triggered, implementation of activity cycles gives rise to actual physical activity, perception, and feeling.

ACT describes and explains the mental processes involved in the continuous sequence of adaptations and interactions that occur in making decisions and choices about how to act in the social and physical worlds. Conscious mental imagery serves a basic adaptive function in enabling a person to prepare, rehearse, and perfect his or her actions. Mental imagery provides the necessary means to guide experimentally and transform experience by running activity cycles as mental simulations of the real thing. Such activity rehearsal can only proceed effectively when the rehearsal incorporates vivid imagery. Imagery that is vivid, through virtue of being as clear and as lively as possible, closely approximates actual perceptual-motor activity, and is of benefit to action preparation, simulation, and rehearsal. In cases where conscious imagery is absent or removed by central nervous system (CNS) injury, there is a need for an alternative means of action planning.

It is established that 2–3% of the population has "congenital aphantasia", meaning that they do not experience visual mental imagery, or lack the ability to control it, but are no less able to control their behavior than people who can visualize. A study of 21 cases of aphantasia found that the majority of participants had some experience of visual imagery from dreams or involuntary "flashes" of imagery, for example, at sleep onset. Thus, their aphantasia involved a deficiency of voluntary imagery rather than a total absence [47]. In one recent study, an aphantasic individual performed significantly worse than controls on the most difficult visual working memory trials [48]. Her performance on a task designed to involve mental imagery did not differ from controls. However, she lacked meta-level cognitive insight into her performance, which is consistent with the current theory. Aphantasic individuals are able to use verbal intentions in goal-directed behavior, schematic spatial imagery, and non-imaged action plans as alternative control systems.

7. Consciousness

After millennia of deep thought, a question that continues to baffle philosophers and psychologists is why humans need consciousness. Knowledge concerning vividness helps answer this question. Mental imagery occurs in a wide range of states of consciousness from waking to sleep. It is the tenet of the current theory that sensory-affective mental images are basic building blocks of consciousness in perception, memory, and imagination [32,44].

Building knowledge requires asking questions. Many times, asking a "good" question leads straight to another question, and so on, until finally there is an answer that may be useful to somebody. No psychological topic prompts more questions than consciousness. When I taught a university BSc Psychology honors course on "consciousness" 40 years ago, it was seen as "off-the-wall" and irrelevant. The only thing was that the students loved this subject and my course received higher ratings that the more traditional courses. Apparently, I was ahead of the game. Now, consciousness is mainstream, and more is known, but there is much more still to learn.

What is consciousness, what is it "made of", and what is it for? To answer these questions, it is sensible to consider what we think we mean when we speak about consciousness and to work from there. I list here 30 claims, indicating which are part of the meta level of executive control (*items in italics*).

(i) *Consciousness is agentic, i.e., it has purpose, desire, and intentionality;*
(ii) *It is deeply social in nature;*
(iii) *It is the center for feelings and moods;*
(iv) *It operates with an inbuilt motivation to drive the organism toward pleasure and away from pain;*
(v) *It is a center for perceptions, interoceptive and exteroceptive;*
(vi) *It serves as a "storehouse" of memories including autobiographical memories from which information and images can be retrieved;*
(vii) *It is the control center for action, perception, attention, affect regulation, cognition, and information processing, all of which require the making of predictions;*
(viii) *It has "layers" and "levels" and is capable of dissociation, splitting, and confusion;*
(ix) *It constructs a personal and a public identity for the "self";*
(x) *It is a center for constructing and changing values and beliefs;*
(xi) *It can set both altruistic and selfish goals, and anything in between;*
(xii) *It can represent information, beliefs, and values in an honest way, or it can simulate, pretend, lie, and be deceitful;*
(xiii) *It can be subject to hearing of voices and other hallucinations;*
(xiv) *It can be subject to illusions and delusions;*
(xv) *It can be accessed by introspection;*
(xvi) *It can be described symbolically in speech, writing, and in works of art, but it can also be ineffable;*
(xvii) *It varies in state of arousal from waking to sleep;*
(xviii) *It references values, beliefs, rules, and customs, and has pragmatic methods for following them;*
(xix) *It strives for the satisfaction of needs including equilibrium;*
(xx) *It can pay close attention to detail or its concentration can wander;*
(xxi) *It fantasizes and "daydreams";*

(xxii) *It plans new goals for the future;*
(xxiii) *It thinks and makes decisions;*
(xxiv) *It imagines and weighs consequences pro and con before acting;*
(xxv) *It receives feedback on the outcomes of action;*
(xxvi) *It "delegates" well-practiced routines, tasks, and habits to a lower level of automatic processing;*
(xxvii) *Automatic functioning, such as the autonomic system, is also below the threshold of consciousness as long as it is performed as expected, but it becomes conscious if it fails to perform normally;*
(xxviii) *It dreams;*
(xxix) *It maintains type II homeostatic responses of the whole organism;*
(xxx) *It remains imperfect.*

Based on the above list, a principle of consciousness (PC) can be stated as follows:

Consciousness is the central executive process of the brain that builds images of the world, makes predictions about future events, and selects which voluntary actions to execute.

The major inputs to consciousness are exteroceptive, sensory stimuli—sight, sound, taste, smell, touch, temperature, vibration, and pain—and also interoceptive stimuli, which form a cortical image of homeostatic afferent activity from the body's tissues. This system provides experiences and visceral feelings such as pain, temperature, itch, sensual touch, muscular and visceral sensations, vasomotor activity, hunger, thirst, and "air hunger". Interoceptive activity is represented in the right anterior insula, providing subjective imagery of the material self as a feeling (sentient) entity, that is, emotional awareness [49]. As the PC states, one of the outputs of consciousness is something that human beings could not possibly do without: predictions. Predictive simulations, otherwise known as rehearsals, involve "what-if" or "if-then" relationships: "If I do X, will Y or Z happen".

Anything that happens between stimulus input and response output is based on if-then operations and simulations geared toward stability and safe prediction. Private fantasies and daydreams take up at least a half of our waking time. It is known that there is a huge quantity of pre-conscious automatic processing of sensory information and behavior that does not require the effortful attention of consciousness. The controlled processing of consciousness is serial, attention-demanding, methodical, and slow, e.g., preparing a meal using a cookery book or reading a manual on how to operate a digital versatile disc (DVD) player. Automatic processing, on the other hand, is efficient and economical, and, quite often, quick, e.g., reading, writing, walking, riding a bicycle, or driving a car.

Brain science supports the idea that the forebrain of the cerebral cortex is the site of the central control system of consciousness. The forebrain itself is involved in regulation of both autonomic and non-autonomic human responses in stress and affect. The forebrain is also the seat of both type I and type II homeostasis.

The significant part of the content of consciousness is mental imagery, the quasi-perceptual mental imagery that gets us from one point on our mental model of the world to the next. Before turning to explore the nature and function of mental imagery, it is essential to say more about how imagery fits into the system as a whole.

8. Psychological Homeostasis

The 19th century French physiologist, Claude Bernard, "the father of modern physiology and experimental medicine", is best known for his work on the pancreas and vasomotor system, and for discovering glycogen. Yet, his description of the "milieu intérieur" in living organisms is equally significant.

"The stability of the internal environment is the condition for the free and independent life."

Bernard's [50] "milieu intérieur" concept was ignored for many decades; nobody really knew what to do with it. Then, in the early 20th Century, J.S. Haldane, C.S. Sherrington, J. Barcroft, and a few others started to work with it. In 1926, the Harvard physiologist Walter Cannon translated the French term into Greek and coined the term homeostasis [51]. Cannon thought that the automatic function of homeostasis freed the brain for more intellectual functions such as intelligence, imagination, insight, and manual skill.

Homeostasis is necessary for every living system and could be the defining characteristic of life itself. At every level of existence, from the cell to the organism, from the individual to the population, and from the local ecosystem to the entire planet, homeostasis is a drive toward stability, security, and adaptation to change. In an infinite variety of forms, omnipresent in living beings, is an inbuilt function with the sole purpose of striving for equilibrium, not only in the "milieu intérieur" but also in the "milieu extérieur". On the other side of Bernard's scientific coin, I postulate the following basic principle:

"The stability of the external environment is the condition for the free and independent life."

By changing a single word "internal" to "external", one creates a whole new theoretical perspective for consciousness, volition, cognition, affect, and behavior, a General Theory of Behavior based on the construct of homeostasis. Striving for balance and equilibrium is the primary guiding force in all that we plan, think, feel, and do. I call this homeostasis (type II) the "reset equilibrium function" (REF) [1].

Every organism automatically regulates essential physiological functions by homeostasis, and internal drives are maintained in equilibrium using corrective behavior in the form of eating, drinking, defecating, sleeping, and so on. This form of homeostasis was established scientifically since the time of Bernard and was implicit as a concept within classical theories of Hippocrates. Far more than this, without any special reflection in most instances, all conscious beings are constantly and quite routinely reconciling the discrepancies among their thoughts, behaviors, and feelings, and in the differences with those with whom they have social relationships. Conscious organisms strive to achieve their goals while maximizing cohesion and cooperation with both kith and kin and, at the same time, strive to take away or minimize the suffering and pain of others. The goal is to minimize all forms of tooth-and-claw competition to live in a culture where the thriving of all is in the self-interest of every individual. This idea was described by Antonio Damasio [9] as "cultural instruments first developed in relation to the homeostatic needs of individuals and of groups as small as nuclear families and tribes. The extension to wider human circles was not and could not have been contemplated. Within wider human circles, cultural groups, countries, and even geopolitical blocs often operate as individual organisms, not as parts of one larger organism, subject to a single homeostatic control. Each uses the respective homeostatic controls to defend the interests of its organism" [9] (p. 32).

Aware of it or not, the REF is omnipresent; wherever one goes and whatever one is doing, the REF is jogging along with us every step of the way. The REF is not something one focuses attention on, but it is nevertheless the process by which our behavioral systems are perpetually striving to maintain balance, safety, and stability in our physical and social surroundings. Competing drives, conflicts, and inconsistencies can all pull the flow of events "off balance", triggering an innate striving to restore equilibrium.

For the majority of time, the majority of people strive to calm and quieten disturbances of equilibrium rather than to acerbate them. It is not a battle that is always won; there is always the possibility of instability, calamity, or catastrophe even. If one cannot win every battle, one can at least strive to win the war.

Courtesy of homeostasis, body and mind are continuously regulating and controlling in multiple domains and levels simultaneously, with constant resets and automatic adjustments to both voluntary and involuntary behavior. Type I homeostasis is the inwardly striving physiological homeostasis H[Φ] and type II homeostasis is the outwardly striving psychological homeostasis H[Ψ]. From birth to death, these two forms of homeostasis provide optimum levels of controllable equilibrium. The reset equilibrium function (REF) integrates the principle of homeostasis with our understanding of psychological processes and behavior. Systems theory with cyclical negative feedback loops is a central feature. Feedback loops in cybernetics and control theory mirror homeostasis within biology and neuroscience. Psychologists employ control theory as a conceptual tool for large areas of psychology [52]. Notably, one objective of control theory was always to provide a "unified theory of human behavior" [53]. A unified theory integrates knowledge about consciousness with knowledge about behavior.

9. The General Theory of Behavior

This General Theory draws upon systems of homeostasis consisting of interconnected processes in continuous feedback loops that are updated with each reset of the REF. The REF extends the reach of homeostasis to a general control function which automatically restores psychological processes to equilibrium and stability. The REF is triggered when any of the processes within a system strays outside of its set point or range. The REF is innate, but it can only exist in conscious organisms which all have two kinds of homeostasis (types I and II). Non-conscious organisms are availed with only one type of homeostasis (Type I). Type II homeostasis exists in a system with any number of processes, each with its own set range, making a series of resets.

Any set of processes, such as the four shown in Figure 6, is a tiny sub-set of thousands of interconnected processes responsible for coding and communicating inside the body and the brain. Any process can be connected to hundreds or thousands of other processes, any one of which can push any particular process out of its "comfort zone", thereby requiring it to reset. As any one process resets, a "domino effect" among many other interconnected processes requires these to reset also. Thus, a reset is often a complete reset of a large part of the entire system, not simply the resetting of a single process. Psychological and physiological processes operate in tandem to maximize equilibrium for each particular set of functions.

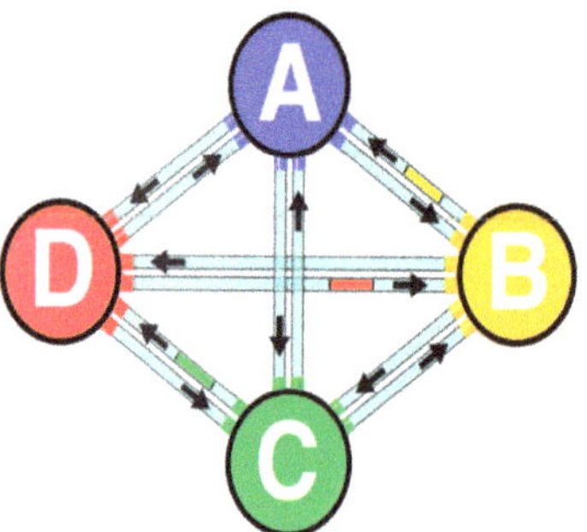

Figure 6. A network of four interconnected processes (A–D) in homeostasis. The reset equilibrium function (REF) returns each process in to its set range. An adjustment in one process may necessitate a compensatory adjustment in one or more of the other interacting processes until equilibrium is reached. Thus, when A stimulates B to lower its activity level, the reduced value in B stimulates C, D, and A.

The General Theory explains the relevance of the REF to numerous psychological functions including those where reset is a condition for change, e.g., affect, chronic stress, excessive behaviors such as smoking, drinking, gambling, and overeating, pain, sleep loss, and low subjective well-being. In all of these situations, the subject's conscious acknowledgement that there is behavior in need of change is of primary importance. This acknowledgement is a necessary precondition for purposeful striving toward making that change.

"The purpose of a brain is not to think, but to act" [54]. The central executive system of consciousness enables organisms to mentally map the environment, predict what might happen next, and to act. One of the major processes for modeling, predicting, and acting is mental imagery. Mental imagery is ideally suited to these purposes by providing preparatory images, which can exist in any sensory modality; however, for the majority of people, this is predominantly visual. On the other hand, imagining the smell and taste of a delicious meal, "hearing" the sound of some enchanting music, and imagining scenes and feelings of relaxation from a recent holiday, or, indeed, "tasting" a delicious glass of wine are all equally possible [55]. Anomalous and paranormal experiences are reported at significantly higher rates among people with vivid VVIQ scores, especially auras, remote healing, and apparitions, but only among high vivid scorers in the open-eyes condition [56].

According to Frederic Bartlett (1932), schemata are much more than elementary reactions ready for use; "they are also arrangements of material, sensory at a low level, affective at a higher level, imaginal at a higher level yet, even ideational and conceptual" [31]. The action system is inextricably linked to the perceptual system, such that perceiving something generally leads to activity in either covert or overt form triggered by schemata. Imagined simulation consists of covert performances in which specific intentions, purposes, and actions are fulfilled [32]. Mentally simulating an experience serves as a substitute for the corresponding experience [57]. Based on this formulation, a mental imagery principle can be stated as follows:

Mental imagery principle: A mental image is a quasi-perceptual experience that includes action schemata, affect, and a goal.

A system based on this principle is shown in Figure 7.

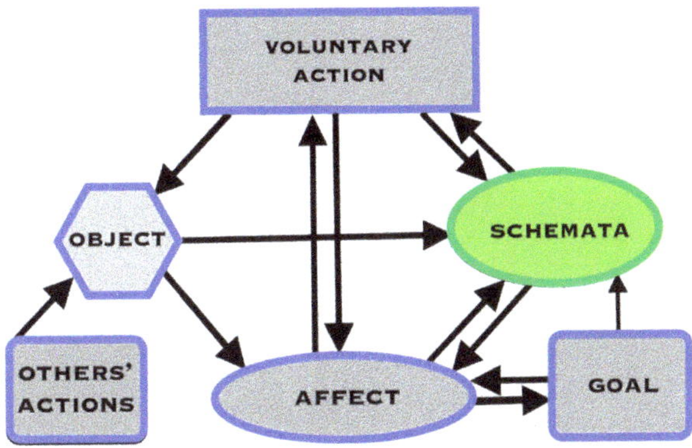

Figure 7. A model of voluntary action ('VOAGA' model). Action schemata (As) control voluntary actions (V) in response to salient objects (O) in the immediate environment, in accordance with current goals (G). Affect (Af) influences the goal and the schemata. Action simulation using mental imagery occurs in the same system as that used for overt action.

Involuntary images, persistent, unpleasant memories, and repetitive habits are symptomatic of disorders, e.g., patients with posttraumatic stress disorder, anxiety disorders, depression, eating disorders, and psychosis frequently report repeated visual intrusions concerning real or imaginary events that can be extremely vivid, detailed, and with highly distressing content [58]. Hallucinations are of particular interest because they are reported by much larger numbers of people than those who have diagnoses of psychosis, and there is a significant overlapping in phenomenology with subjective paranormal experience such as precognition and out-of-the-body experience. One definition of hallucination states it to be "any percept-like experience which (a) occurs in the absence of an appropriate stimulus, (b) has the full force of impact of the corresponding actual (real) perception, and (c) is not amenable to the direct or voluntary control of the experiencer [59] (p. 23). In the first two parts, hallucination is a form of mental imagery. The third part provides the distinguishing feature because, to be beneficial, mental images need to be voluntary and controllable by the experiencer. The full spectrum of conscious experience, including dissociative states, psychotic episodes, hallucination, pseudo-hallucinations, out-of-the-body experiences, delusions, ipseity, subjective paranormal experience, and the varying ability to exert voluntary control, is the subject of another paper [60].

As noted, the General Theory proposes a cyclical system of objects, schemata, affective experience, and actions. The control system has both an executive level and a schema level. The executive level, which is what is normally referred to as "consciousness", controls and monitors the schema level. This duality of levels enables moment-by-moment adjustments to goal-seeking behavior at the schema level. Goals are set at the executive level of consciousness. Goal-setting is guided by values and beliefs which inform actions, inhibit actions, or reflect on what action to take, as the situation requires.

In competent performers, speech, decisions, routines, and many complex behaviors normally do not require conscious control to operate [61]. Afferents from the muscles and the activity of the cerebellum, where movement is organized, operate entirely preconsciously and produce no conscious images [62,63].

Conscious imagery is useful in the planning and organization of behavior through enabling the simulation of action sequences at the object level without energy expenditure or risk. The object level interfaces with the social level in the public domain of shared activities and object levels. The possible outcomes of alternative future actions can be appraised prior to a course of action. In this way, conscious mental imagery serves as a mental toolbox, producing its internal contents for the user to explore and manipulate in the selection and preparation of future physical and social activity.

10. The Clock System

An internal clock controls physiological and behavioral processes of daily living in synchrony with regular changes in the environment. Over hundreds of millions of years in an environment that changes dramatically over every 24-h cycle, evolution produced universal rhythms throughout the plant and animal kingdoms such that each organism's biochemistry, physiology, and behavior are organized in diurnal cycles [64]. Many circadian rhythms are persistent even in the absence of the normal diurnal cues of night and day or temperature changes, e.g., while living in caves. Such demonstrations are interpreted as reflecting the operation of an internal biological clock or clocks. The circadian clock system serves as a biological "alert" that lets us know when significant events are due to happen.

The light-dark (LD) cycle is the most reliable of the external signals enabling entrainment and is referred to as a "zeitgeber" (i.e., time-giver). LD information is perceived by mammals with retinal photoreceptors and conveyed directly to the suprachiasmatic nucleus (SCN) of the hypothalamus, where it entrains oscillators in what is regarded as the master clock of the organism [65]. Other cyclic inputs, such as temperature, noise, social cues, or fixed mealtimes, also can act as entraining and predictive agents, although usually to a less reliable extent than LD.

An entrainable circadian clock is present in the SCN during fetal development, and the maternal circadian system coordinates the phase of the fetal clock to environmental lighting conditions. Even before birth, the organism is entrained to the LD cycle [66]. Having a clock system is advantageous for predicting and preparing for important events. When food is available only for a limited time each day, it was observed that rats increase their locomotor activity two to four hours before the onset of food availability [67]. Similar anticipatory behavior occurs in other mammals and in birds, accompanied by increases in body temperature, adrenal secretion of corticosterone, gastrointestinal motility, and activity of digestive enzymes.

It was proposed that a common design principle applies to the clock in all organisms, from bacteria to humans, and that the circadian clock existed for at least 2.5 billion years. [64]. The predictive mechanism in which physiology and behavior are "tuned" to the timing of external events allows a competitive advantage. When disrupted by genetic or environmental means, cardio-metabolic diseases and cancer can be triggered, and realigning out-of-sync circadian rhythms can be beneficial in the treatment of endocrine-related disorders [67].

Synchronicity at a neural level with 35–75-Hz oscillations in the cerebral cortex, hypothesized to be "the basis of" consciousness, form the binding that may be achieved by the synchronized oscillations of neuronal groups [68]. It is suggested that two items of information (e.g., shape and color) are bound together if the relevant neural groups oscillate with the same frequency and phase. Consciousness, I would humbly suggest, is an emergent property of evolution.

11. The Approach Avoidance Inhibition (AAI) System

"Every person on the planet (barring illness) can tell good from bad, positive from negative, pleasure from displeasure" [69]. Not only can one tell it, one can feel it also. From the pre-Socratic philosophers until the present day, the role of pleasure and pain as motivators of human behavior is universally accepted. Psychological hedonism, the idea that all action is determined by the degree of pleasure or displeasure that imagining the action provokes, dates back to Epicurus (341–270 BC)), who is alleged to have said "we begin every act of choice and avoidance from pleasure ... " [70]. The idea that organisms strive for pleasure and the avoidance of pain was accepted for eons. Michel Cabanac suggests that the pleasure or displeasure of a sensation is directly related to the biological usefulness of the stimulus [71]. The seeking of pleasure and the avoidance of displeasure have useful homeostatic consequences. That is, they depend on the internal state of the stimulated subject at the particular moment of the stimulation. Pleasure indicates a useful stimulus and motivates approach, while pain indicates a useful stimulus and motivates avoidance.

Emerging evidence indicates similarities in the anatomical substrates of painful and pleasant sensations in the opioid and dopamine systems [72]. The experiences of positive and negative affect are based on neural circuits that evolved to ensure survival. These circuits are activated by external stimuli that are appetitive and life-sustaining or by stimuli that threaten survival. Activation of the pain and pleasure circuits alert the sensory systems to pay attention and prompt motor action [72]. The approach-avoidance concept was pivotal in theories of behavior [73]. The approach-avoidance system includes behavioral inhibition which takes over when there is approach-avoidance conflict. The approach-avoidance-inhibition (AAI) system sets the bar for fight-fright-freeze decisions that are pervasive throughout the animal kingdom. Action schemata are necessary precursors to action in a four-pronged system for regulating approach-avoidance-inhibition (AAIS). Operating together with action schemata, the REF, clock, and AAIS regulate voluntary action (Figure 8).

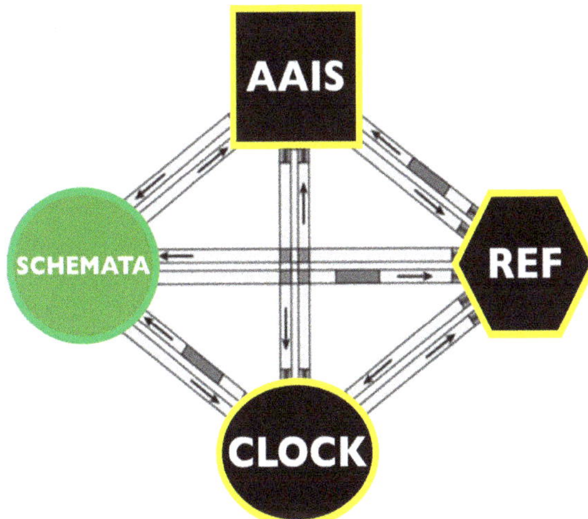

Figure 8. The reset equilibrium function (REF), clock, and approach-avoidance-inhibition system (AAIS) interconnect with action schemata to regulate voluntary action.

In the following section, the different modular systems described for imagery, timing, action, and inhibition are integrated into a single system for behavior control.

12. Behavior Control System

As the executive controller of brain and behavior, consciousness occupies pole position, providing a significant evolutionary advantage to the organism. Figure 9 shows the behavior control system for the planning and execution of behavior. The system for the regulation of emotion [74] is confined here to a single module for affect. It is accepted that consciousness is the coordinated activity of the brain system as a whole, a direct emergent property of cerebral activity [75]. Consciousness exerts supervening control over the entire flow pattern of cerebral excitation. The system shows large modular functional entities with a huge range of special qualities and properties. The modules interact causally with one another as homeostatic entities to produce the entire gamut of images, feelings, thoughts, and actions, all working toward stability and equilibrium. Mental imagery is a benefit to the planning and prediction roles of consciousness, to the function of psychological homeostasis. If vividness is entirely lacking, then language provides an alternative route (e.g., "first do X, then do Y"), such that, by whatever means, psychological homeostasis can ensure that equilibrium with the environment is maximized at all times.

Critics claim that consciousness is an epiphenomenon and that its contents are formed "backstage" by non-conscious systems. Oakley and Halligan [76] stated that "psychological processing and psychological products are not under the control of consciousness ... All 'contents of consciousness' are generated by and within non-conscious brain systems in the form of a continuous self-referential personal narrative that is not directed or influenced in any way by the 'experience of consciousness'". The behavior control system contains a meta level, a schema level, and an automatized level. It is the executive meta level of consciousness that sets goals and directs the lower levels to prepare and execute actions, and make necessary adjustments. Conscious mental images are also put into service at the schema level, e.g., in simulation, skills, design, drawing, writing, geometry, and other creative performance. The majority of behavior, which can be classed as routine, falls within the automatized level and does not require consciousness. The behavior control system enables decisions about outcomes of alternative future actions to be weighed and appraised prior to committing to any course of action. In addition to language skills, conscious imagery serves as a mental "toolbox", providing its internal contents for the user to select and prepare future physical and social activity. It is the meta level of consciousness, having the quality of ipseity, an awareness of a unique personal, meaningful point of view, driven by values, beliefs, feelings, desires, and wants, by the universal striving of psychological homeostasis, that sets new projects, monitors progress, and directs the schema level to prepare and execute actions.

According to a personal narrative account [76], the experience of consciousness is telling stories to ourselves about a fictive internal state of "consciousness". I have strong doubts about this account because it leaves so much unexplained. How does this account explain the experience of pain [77]; why must researchers be given ethical guidelines for investigations of experimental pain in conscious animals? [78]; why does the science of psychology require the concept of dreams [79] and hallucinations? [80]; how are thought experiments employed to create scientific theories? [81]; why are there objective performance gains from mental rehearsal by elite athletes, footballers, and others? [82]; why should differences in visual perspectives, "external" (third-person) and "internal" (first-person), be differentially interfered with by a concurrent action dual task? [83]; why, in reading poems, "vividness of imagery was the strongest contributor to aesthetic pleasure, followed by valence and arousal"? [14]; why is vividness of mental images associated with a stronger sense of presence felt in experiencing virtual reality scenarios? [84]; and why do visualization, first-person perspective, and narratives representing real experiences improve memory and comprehension? [85]. Dozens more examples of how consciousness acts as a meta-level controller of behavior are available. There would be far too many mysteries left to explain if consciousness was simply a figment of a personal narrative.

Another argument against the view that consciousness controls action and thought is the empirical evidence that the outcome of a decision can be encoded in brain activity of prefrontal and parietal cortex seconds before it enters awareness. It is presumed that this delay reflects the operation of a network of high-level control areas that begin to prepare an upcoming decision long before it enters awareness [86]. We need to look no further for an explanation than to Libet himself: "Although the volitional process may be initiated by unconscious cerebral activities, conscious control of the actual motor performance of voluntary acts definitely remains possible. The findings should, therefore, be taken not as being antagonistic to free will, but rather as affecting the view of how free will might operate. Processes associated with individual responsibility and free will would 'operate' not to initiate a voluntary act but to select and control volitional outcomes." [86] (p. 538). My General Theory specifies a level of operations for automatized actions such as tying one's shoe laces, riding a bicycle, or visiting the bathroom, a category within which Libet's task clearly falls (Figure 9). The last thing consciousness needs is to be aware of every single little detail in one's thoughts, actions, and feelings. What it definitely must have at its disposal is the personal self, with purposes, desires, and intentions. There were many criticisms of the original studies by Libet [86]. As Bridgeman, the first of multiple commentators on the Libet study, pointed out, "a careful analysis of the experimental conditions reveals that the subjects' wills were not as free as the Libet article implies, for the small, sharp movements that they were instructed to make were not freely willed but were requested by the experimenter. The will of a subject was no more free in this design than in reaction-time experiments; the only difference between this experiment and the latter paradigms is that the instruction and the movement are decoupled in time. While performing the task, the subjects do nothing more than obey the instructions" [87] (p. 540).

The authors of another Libet-style readiness potential study [88] suggested "free will is an illusion" because they found evidence of high-level control areas beginning to determine an upcoming decision 10 s before entering awareness. The task consisted of pushing a button using an index finger under a prescription by the investigators to choose "freely". The task was basic, routine, automatized, and under the control of another person's demands. A free human operator would be "tied up in knots" if all the minute details of such an elementary task needed conscious, volitional control. Nature was kind when it took such automatized actions out of conscious control. Mechanisms for automatic control mechanisms are on a different level to volitional control. For example, when picking up an object, one tends to grasp at contact points that allow a stable grip. Appropriate grasp points can be re-selected during an ongoing movement in response to unexpected perturbations of the target object, suggesting that automatic control mechanisms guide the fingers to appropriate grasp points distinctly from those involved with volitional control [89]. Volitional control with meta-level consciousness is only required when a new purpose for a new project is formulated. The key point is that new project has a personal meaning and coherence within the life-space of a unique individual. Consciousness is required only when skilled actions are to be executed, actions that might interfere with actions already ongoing, including those of others (Figure 9). Feeling, meaning, intention, and a sense of timing and flow are important influences on the quality and energy of actions. Feeling is integrated with the meaning and purpose of an action through the meta system of consciousness.

Some critics claim that totally unconscious living beings could conceivably exhibit all of the adaptive behaviors of conscious beings, the so-called "zombie" option. They suggest that it is entirely conceivable and coherent to state that all structural and functional characteristics of living beings could be explained without any accompanying conscious experience. It is metaphysically possible that such unconscious zombies could be "in the dark" and be equally effective. This is the "zombie" hypothesis of conscious inessentialism in the philosophy of the mind [90]. The zombie is a fascinating hypothetical concept and nobody has ever yet claimed to know, be or to have built a zombie. If so, it would be suitably undone. The acid test would be to give it/her/him an fMRI while answering the items of the VVIQ, or to examine the zombie's electroencephalogram (EEG) and eye movements during sleep. Any mutterings about images or dreams could be quickly disposed of when the brain scans failed to show the requisite patterns of activation. The zombie is a freak, never to be found in Nature.

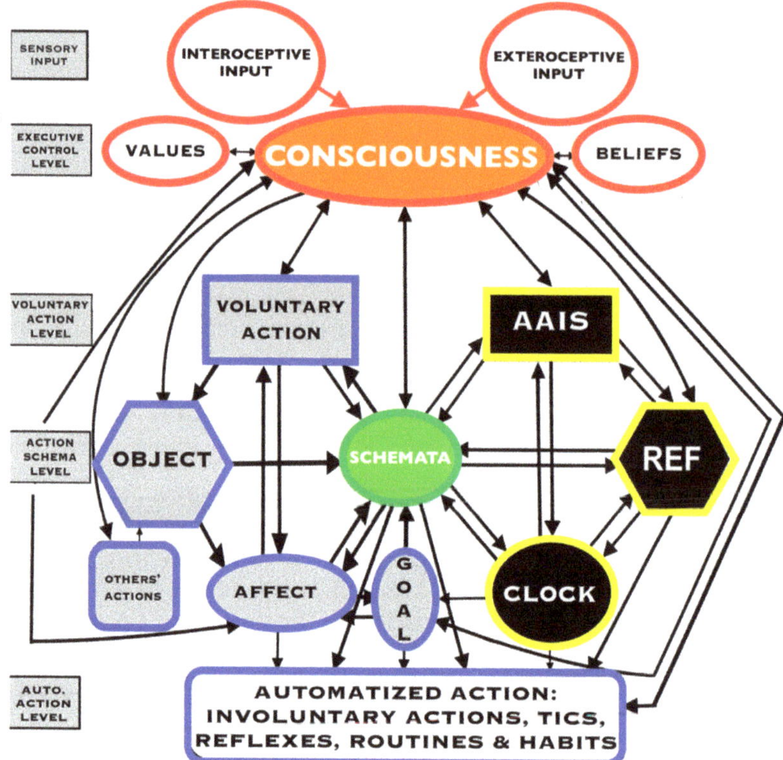

Figure 9. The behavior control system shown with five levels of organization and nine modular systems for the generation of action with consciousness at the executive level. The modules fall into three groups: (i) the ACT group of modules from Figure 7 (gray and blue); (ii) the REF (type II homeostasis), clock (circadian system), and AAIS (approach/avoidance/inhibition system) from Figure 8 (black and yellow) interconnecting with the ACT group via action schemata (green); (iii) the behavior control system/consciousness (red, white, and orange) includes modules for interoceptive and exteroceptive sensory input, beliefs, and values. Five levels of control are sensory input, executive control, voluntary behavior, the AAIS, action schemata, and REF, and the automatized action level. The structure of the affective system (not shown) is mediated by the limbic system, a lateral circuit passing through the hypothalamus regulating internal and hormonal processes, the cingulate cortex, the hippocampus, and amygdala.

Why did consciousness evolve? I suggest the answer lies in the significant evolutionary advantages of an emergent, purposeful, and integrative process of psychological homeostasis to direct holistic control [9,75]. The case for consciousness as an emergent property was made by Roger Sperry in 1969:

> "The long-standing assumption in the neurosciences that the subjective phenomena of conscious experiences do not exert any causal influence on the sequence of events in the physical brain process is directly challenged in this current view of the nature of mind and the mind-brain relationship. A theory of mind is suggested in which consciousness, interpreted to be a direct emergent property of cerebral activity, is conceived to be an integral component

of the brain process that functions as an essential constituent of the action and exerts a directive holistic form of control over the flow pattern of cerebral excitation" [75] (p. 532)

Consciousness has an explanatory role in the behavior of organisms through its equilibrium-creating process of meta-level control. An experience that is re-represented in the mind has conscious meta cognition, or self-reflection. The intimate connection between the self, intentionality, and purposive goal-seeking behavior has the pre-eminent advantage of ipseity. Only a conscious being can have the invaluable quality of selfhood, the implicit first-person quality of consciousness that all experience articulates from a first-person perspective as "*my*" experience [91]. The zombie can know nothing of this.

13. Homeostasis as a Unifying Concept

Homeostasis is a unifying concept across the disciplines of consciousness studies cognitive neuroscience, psychology, and biology. Yet, it is a much neglected concept. Recent studies of the brain's anatomical connectivity or "connectome" show that the CNS is at once more complex and more simple than previously assumed. Regions of interest produce coherent fluctuations in neural activity and distributed patterns of activation or networks. Neurobiological networks occur at different organizational levels from cell-specific regulatory pathways inside neurons to interactions between systems of cortical areas and subcortical nuclei. Architectures which support cognition, affect, and action are normally found at the highest level of analysis [92]. Brian Edlow and colleagues investigated the limbic and forebrain structures that form a "central homeostatic network" (CHN) [93] responsible for autonomic, respiratory, neuroendocrine, emotional, immune, and cognitive adaptations to stress. These structures include the limbic system in shared participation in homeostasis. Recent research focused on homeostatic forebrain nodes which receive sensory information concerning extrinsic threats and intrinsic metabolic derangements from the brainstem, resulting in arousal from sleep, heightened attention, vigilance during waking, and visceral and somatic motor defenses. These findings suggest that homeostasis is mediated by ascending and descending interconnections between brainstem nuclei and forebrain regions, which together regulate autonomic, respiratory, and arousal responses to stress. The role of the limbic system in the regulation of homeostasis is being recognized, and the limbic system was added to the central autonomic network of "flight, fight, or freeze". These findings suggest that homeostasis of type I H[Φ] and type II H[Ψ] are controlled by a single executive controller in the forebrain.

That the forebrain controls both types of homeostasis supports the contention that homeostasis is a unifying concept across biology and psychology. Studies of aging suggest that homeostatic dysregulation proceeds in parallel in multiple physiological systems [94]. Aging is characterized by marked reductions in functional correlations within higher-order brain systems [95]. Physiological and psychological homeostasis can be modeled in the same way as a system of adjustments through a network of connected processes or states [1].

14. Conclusions

Consciousness is an open system having many relations to its mental, physical, and social surroundings. Changes in these surroundings produce internal "disturbances" of the system that require adjustment, adaptation, or correction. As originally described by Bernard [50] and Cannon [51], and more recently by the author [1], such disturbances are normally kept within set limits, because automatic adjustments are brought into action such that the internal and external conditions are held fairly constant. Everything known about the executive role of the forebrain in action planning and decision-making and the recently discovered central homeostatic network [93] suggests that this must indeed be the case.

The General Theory of Behavior holds that the reach of homeostasis extends well beyond physiology into many realms of psychology and into society as a whole. Homeostasis type I, H[Φ], and type II, H[Ψ], serve identical stabilizing functions internally in the body and externally in socio-physical interactions, respectively. With Cannon, I hypothesize that "steady states in society as a whole and steady states in its members are closely linked" [51]. H[Φ] and H[Ψ] exist in a complementary relationship of mutual support.

A much neglected topic in the history of psychology and today in consciousness studies is mental imagery. Yet, the evidence suggests that mental imagery is the basic building block of consciousness [33]. In protecting stability, security, and equilibrium in the socio-physical world, psychological homeostasis is one of the primary functions of consciousness, and could not exist without it. The emergence of homeostasis and consciousness in evolution would not have been possible without mental imagery. It is impossible to say which came first, consciousness or psychological homeostasis. One thing of which we can be certain is that each of us can truthfully say: "I am conscious; therefore, I am".

Funding: This research received no external funding.

Acknowledgments: I am pleased to acknowledge a debt of gratitude for critique, information, or assistance to many people with whom I have had the good fortune of meeting, collaborating with, and, in many instances, becoming friends: Akhter Ahsen, Alan Richardson, Allan Paivio, Anne Isaac, Carole Ernest, Cesare Cornoldi, Donald Broadbent, Doris McIlwain, Elizabeth Loftus, Geoff Loftus, Geir Kaufmann, George Sperling, Gosaku Naruse, Ian Hunter, Jack Clarkson, Jack Hilgard, John Richardson, Joseph Juhasz, Mark Marschark, Martin Orne, Michel Denis, Peter Hampson, Peter McKellar, Peter Sheehan, Ray Hyman, Robert Morris, Stanley Krippner, Shinsuke Hishitani, Steve Kosslyn, Stuart McKelvie, Takeo Hatakeyama, and Tess Molteno. Graham McPhee kindly provided a forerunner of Figure 6. In addition, I appreciate the helpful comments of the two reviewerswho enabled me to improve the clarity of the paper. Responsibility for error rests entirely with the author.

Conflicts of Interest: The author declares no conflict of interest.

References

1. Marks, D.F. *A General Theory of Behaviour*; SAGE Publications: London, UK, 2018.
2. Betts, G.H. *The Distribution and Functions of Mental Imagery*; Columbia University: New York, NY, USA, 1909.
3. Sheehan, P.W. A shortened form of Betts' questionnaire upon mental imagery. *J. Clin. Psychol.* **1967**, *23*, 386–389. [CrossRef]
4. Marks, D.F. Individual Differences in the Vividness of Visual Imagery and their Effect on Function. In *The Function and Nature of Imagery*; Sheehan, P.W., Ed.; Academic Press: New York, NY, USA, 1972; pp. 83–108.
5. Runge, M.S.; Cheung, M.W.-L.; D'Angiulli, A. Meta-analytic Comparison of trial-versus questionnaire-based vividness reportability across behavioral, cognitive and neural measurements of imagery. *Neurosci. Conscious.* **2017**, *1*, nix006.
6. Kaplan, A. *The Conduct of Inquiry*; Chandler: San Francisco, CA, USA, 1964.
7. Kind, A. Imaginative Vividness. *J. Am. Philos. Assoc.* **2017**, *3*, 32–50. [CrossRef]
8. Holmes, E.A.; Mathews, A. Mental Imagery and Emotion: A Special Relationship? *Emotion* **2005**, *5*, 489–497. [CrossRef] [PubMed]
9. Damasio, A. *The Strange Order of Things: Life, Feeling, and the Making of Cultures*; Vintage: New York, NY, USA, 2018; p. 32.
10. Pylyshyn, Z.W. What the Mind's Eye Tells the Mind's Brain: A Critique of Mental Imagery. *Images Percept. Knowl.* **1973**, *80*, 1–36.
11. Isaac, A.; Marks, D.F.; Russell, D.G. An instrument for assessing imagery of movement: The Vividness of Movement Imagery Questionnaire (VMIQ). *J. Ment. Imagery* **1986**, *10*, 23–30.
12. Hebb, D. Concerning Imagery. *Psychol. Rev.* **1968**, *75*, 466–477. [CrossRef]
13. Lang, P.J. A Bio-Informational Theory of Emotional Imagery. *Psychophysiology* **1979**, *16*, 495–512. [CrossRef]
14. Belfi, A.M.; Vessel, E.A.; Starr, G.G. Individual ratings of vividness predict aesthetic appeal in poetry. *Psychol. Aesthet. Creat. Arts* **2018**, *12*, 341–350. [CrossRef]
15. Marks, D.F.; McKellar, P. The Nature and Function of Eidetic Imagery. *J. Ment. Imagery* **1982**, *6*, 1–124.
16. Richardson, A. Mental Practice: A Review and Discussion Part I. *Q. Am. Assoc. Heal. Phys. Educ.* **1967**, *38*, 95–107. [CrossRef]
17. Jeannerod, M. Neural Simulation of Action: A Unifying Mechanism for Motor Cognition. *NeuroImage* **2001**, *14*, S103–S109. [CrossRef]

18. Feltz, D.L.; Landers, D.M. The Effects of Mental Practice on Motor Skill Learning and Performance: A Meta-analysis. *J. Sport Psychol.* **1983**, *5*, 25–57. [CrossRef]
19. Markman, K.D.; Klein, W.M.P.; Suhr, J.A. *Handbook of Imagination and Mental Simulation*; Psychology Press: New York, NY, USA, 2009.
20. Marks, D.F.; Isaac, A.R. Topographical distribution of EEG activity accompanying visual and motor imagery in vivid and non-vivid imagers. *Br. J. Psychol.* **1995**, *86*, 271–282. [CrossRef]
21. Isaac, A.R.; Marks, D.F. Individual differences in mental imagery experience: Developmental changes and specialization. *Br. J. Psychol.* **1994**, *85*, 479–500. [CrossRef]
22. Pichiorri, F.; Morone, G.; Petti, M.; Toppi, J.; Pisotta, I.; Molinari, M.; Paolucci, S.; Inghilleri, M.; Astolfi, L.; Cincotti, F.; et al. Brain-computer interface boosts motor imagery practice during stroke recovery. *Ann. Neurol.* **2015**, *77*, 851–865. [CrossRef] [PubMed]
23. Hegarty, M. Mechanical reasoning by mental simulation. *Trends Cogn. Sci.* **2004**, *8*, 280–285. [CrossRef] [PubMed]
24. Marks, D.F. Visual imagery differences in the recall of pictures. *Br. J. Psychol.* **1973**, *64*, 17–24. [CrossRef] [PubMed]
25. McKelvie, S.J. The VVIQ as a psychometric test of individual differences in visual imagery vividness: A critical quantitative review and plea for direction. *J. Ment. Imagery* **1995**, *19*, 1–106.
26. Cui, X.; Jeter, C.B.; Yang, D.; Montague, P.R.; Eagleman, D.M. Vividness of mental imagery: Individual variability can be measured objectively. *Vis. Res.* **2007**, *47*, 474–478. [CrossRef]
27. Dijkstra, N.; Bosch, S.E.; Van Gerven, M.A. Vividness of Visual Imagery Depends on the Neural Overlap with Perception in Visual Areas. *J. Neurosci.* **2017**, *37*, 1367–1373. [CrossRef] [PubMed]
28. Rooney, W. FourFourTwo. Performance. 2019. Available online: https://www.fourfourtwo.com/us/performance/training/wayne-rooney-big-match-preparation (accessed on 9 May 2019).
29. Paivio, A. *Imagery and Verbal Processes*; Psychology Press: New York, NY, USA, 2013.
30. Ganis, G.; Thompson, W.L.; Kosslyn, S.M. Brain areas underlying visual mental imagery and visual perception: An fMRI study. *Cogn. Brain* **2004**, *20*, 226–241. [CrossRef]
31. Bartlett, F.C. *Remembering*; Cambridge University Press: Cambridge, UK, 1932.
32. Marks, D.F. On the Relationship Between Imagery, Body and Mind. In *Imagery: Current Developments*; Hampson, P.J., Marks, D.F., Richardson, J.T.E., Eds.; Routledge: New York, NY, USA, 1990; pp. 1–38.
33. Marks, D.F. Consciousness, mental imagery and action. *Br. J. Psychol.* **1999**, *90*, 567–585. [CrossRef]
34. Wundt, W.M. *Outlines of Psychology*; Wilhelm Engelmann: Leipzig, Germany, 1907.
35. Tomkins, S.S. *Affect Imagery Consciousness: Volume I: The Positive Affects*; Springer Publishing Company: New York, NY, USA, 1962.
36. Zajonc, R.B. Feeling and thinking: Preferences need no inferences. *Am. Psychol.* **1980**, *35*, 151–175. [CrossRef]
37. Kinzel, V.; Kubler, D. Facial Muscle Patterning to Affective in Depressed and Nondepressed Subjects. *Proc. Natl. Acad. Sci. USA* **1971**, *68*, 2153.
38. Barrett, L.; Bar, M. See it with feeling: Affective predictions during object perception. *Philos. Trans. Soc. B Boil. Sci.* **2009**, *364*, 1325–1334. [CrossRef] [PubMed]
39. Baars, B.J. *Theatre of Consciousness: The Workspace of the Mind*; Oxford University Press: New York, NY, USA, 1997.
40. Jaynes, J. *The Origins of Consciousness in the Breakdown of the Bicameral Mind*; Houghton Mifflin Harcourt: New York, NY, USA, 2000.
41. Wulf, G.; McNevin, N.; Shea, C.H. The automaticity of complex motor skill learning as a function of attentional focus. *Q. J. Exp. Psychol. Sect. A* **2001**, *54*, 1143–1154. [CrossRef]
42. James, W. *The Principles of Psychology*; Holt and Company: New York, NY, USA, 1890.
43. Jeannerod, M. The 25th Bartlett Lecture: To act or not to act: Perspectives on the representation of actions. *Q. J. Exp. Psychol. A* **1999**, *52*, 1–29. [CrossRef] [PubMed]
44. Marks, D.F. The neuropsychology of imagery. In *Theories of Image Formation*; Marks, D.F., Ed.; Brandon House: New York, NY, USA, 1986; pp. 225–242.
45. Feinberg, T.E.; Mallatt, J.M. *The Ancient Origins of Consciousness*; MIT Press: Boston, MA, USA, 2016.
46. Bacon, F. *The Advancement of Learning*; Cassell & Company: London, UK, 1605.
47. Zeman, A.; Dewar, M.; Della Sala, S. Reflections on aphantasia. *Cortex* **2016**, *74*, 336–337. [CrossRef] [PubMed]

48. Jacobs, C.; Schwarzkopf, D.S.; Silvanto, J. Visual working memory performance in aphantasia. *Cortex* **2018**, *105*, 61–73. [CrossRef] [PubMed]
49. Craig, A. Interoception: The sense of the physiological condition of the body. *Curr. Opin. Neurobiol.* **2003**, *13*, 500–505. [CrossRef]
50. Bernard, C. Lectures on the phenomena common to animals and plants trans. In H. Hoff, R. Guillemin, and L. Guillemin (review). *Perspect. Biol. Med.* **1975**, *19*, 151–152.
51. Cannon, W.B. Organization for physiological homeostasis. *Physiol. Rev.* **1929**, *9*, 399–431. [CrossRef]
52. Carver, C.S.; Scheier, M.F.; Weintraub, J.K. Assessing coping strategies: A theoretically based approach. *J. Pers. Soc. Psychol.* **1989**, *56*, 267–283. [CrossRef]
53. Grinker, R.R.; Hughes, H.M. *Towards a Unified Theory of Human Behaviour*; Basic Books: New York, NY, USA, 1956.
54. Kunz, E. Henri Laborit and the inhibition of action. *Dialogues Clin. Neurosci.* **2014**, *16*, 113–117.
55. Croijmans, I.; Speed, L.J.; Arshamian, A.; Majid, A. Measuring Multisensory Imagery of Wine: The Vividness of Wine Imagery Questionnaire. *Multisensory Res.* **2019**, *1*, 1–17. [CrossRef]
56. Parra, A. Seeing rare things with the mind's eye: Visual imagery vividness and paranormal/anomalous experiences. *Aust. J. Parapsychol.* **2015**, *15*, 37–51.
57. Kappes, H.B.; Morewedge, C.K. Mental Simulation as Substitute for Experience. *Soc. Pers. Psychol. Compass* **2016**, *10*, 405–420. [CrossRef]
58. Brewin, C.R.; Gregory, J.D.; Lipton, M.; Burgess, N. Intrusive images in psychological disorders: Characteristics, neural mechanisms, and treatment implications. *Psychol. Rev.* **2010**, *117*, 210. [CrossRef]
59. Slade, P.D.; Bentall, R.P. *Sensory Deception: A Scientific Analysis of Hallucination*; Johns Hopkins University Press: Baltimore, MD, USA, 1988.
60. Marks, D.F. The spectrum of conscious experience. Unpublished work. 2019.
61. Bargh, J.A.; Ferguson, M.J. Beyond behaviorism: On the automaticity of higher mental processes. *Psychol. Bull.* **2000**, *126*, 925–945. [CrossRef]
62. Schmahmann, J.D.; Sherman, J.C. The cerebellar cognitive affective syndrome. *Brain* **1998**, *121*, 561–579. [CrossRef] [PubMed]
63. Guell, X.; Gabrieli, J.D.; Schmahmann, J.D. Triple representation of language, working memory, social and emotion processing in the cerebellum: Convergent evidence from task and seed-based resting-state fMRI analyses in a single large cohort. *NeuroImage* **2018**, *172*, 437–449. [CrossRef]
64. Loudon, A.S. Circadian Biology: A 2.5 Billion Year Old Clock. *Curr. Boil.* **2012**, *22*, R570–R571. [CrossRef]
65. Stokkan, K.-A. Entrainment of the Circadian Clock in the Liver by Feeding. *Science* **2001**, *291*, 490–493. [CrossRef] [PubMed]
66. Reppert, S.M.; Schwartz, W.J.; Pearson, J.F. Maternal Coordination of the Fetal Biological Clock in Utero. *Obstet. Anesthesia* **1983**, *3*, 113. [CrossRef]
67. Mistlberger, R.E.; Rechtschaffen, A. Recovery of anticipatory activity to restricted feeding in rats with ventromedial hypothalamic lesions. *Physiol. Behav.* **1984**, *33*, 227–235. [CrossRef]
68. Gamble, K.L.; Berry, R.; Frank, S.J.; Young, M.E. Circadian clock control of endocrine factors. *Nat. Rev. Endocrinol.* **2014**, *10*, 466–475. [CrossRef] [PubMed]
69. Lindquist, K.A.; Satpute, A.B.; Wager, T.D.; Weber, J.; Barrett, L.F. The Brain Basis of Positive and Negative Affect: Evidence from a Meta-Analysis of the Human Neuroimaging Literature. *Cereb. Cortex* **2015**, *26*, 1910–1922. [CrossRef]
70. Calne, D.B. *Within Reason: Rationality and Human Behavior*; Vintage: New York, NY, USA, 2000.
71. Clements-Croome, D.; Cabanac, M. Pleasure and joy, and their role in human life. In *Creating the Productive Workplace*; CRC Press: Boca Raton, FA, USA, 1999; pp. 40–50.
72. Lang, P.J.; Bradley, M.M. Emotion and the motivational brain. *Biol. Psychol.* **2010**, *84*, 437–450. [CrossRef]
73. Elliot, A.J. Approach and avoidance motivation and achievement goals. *Educ. Psychol.* **1999**, *34*, 169–189. [CrossRef]
74. Etkin, A.; Büchel, C.; Gross, J.J. The neural bases of emotion regulation. *Nat. Rev. Neurosci.* **2015**, *16*, 693–700. [CrossRef]
75. Sperry, R.W. A modified concept of consciousness. *Psychol. Rev.* **1969**, *76*, 532–536. [CrossRef]
76. Oakley, D.A.; Halligan, P.W. Chasing the Rainbow: The Non-conscious Nature of Being. *Front. Psychol.* **2017**, *8*, 1924. [CrossRef]

77. Melzack, R. The McGill Pain Questionnaire: Major properties and scoring methods. *PAIN* **1975**, *1*, 277–299. [CrossRef]
78. Zimmermann, M. Ethical guidelines for investigations of experimental pain in conscious animals. *PAIN* **1983**, *16*, 109–110. [CrossRef]
79. Freud, S.; Strachey, J. *The Interpretation of Dreams*; Macmillan: London, UK, 1916; p. 311526.
80. Silbersweig, D.A.; Stern, E.; Frith, C.; Cahill, C.; Holmes, A.; Grootoonk, S.; Seaward, J.; McKenna, P.; Chua, S.E.; Schnorr, L.; et al. A functional neuroanatomy of hallucinations in schizophrenia. *Nat. Cell Boil.* **1995**, *378*, 176–179. [CrossRef] [PubMed]
81. Holton, G.J. *Thematic Origins of Scientific Thought: Kepler to Einstein*; Harvard University Press: Cambridge, MA, USA, 1988.
82. Driskell, J.E.; Copper, C.; Moran, A. Does mental practice enhance performance? *J. Appl. Psychol.* **1994**, *79*, 481–492. [CrossRef]
83. Callow, N.; Edwards, M.G.; Jones, A.L.; Hardy, L.; Connell, S. Action dual tasks reveal differential effects of visual imagery perspectives on motor performance. *Q. J. Exp. Psychol.* **2018**, *1747021818811464*. [CrossRef] [PubMed]
84. Iachini, T.; Maffei, L.; Masullo, M.; Senese, V.P.; Rapuano, M.; Pascale, A.; Sorrentino, F.; Ruggiero, G. The experience of virtual reality: Are individual differences in mental imagery associated with sense of presence? *Cogn. Process.* **2018**, 1–8. [CrossRef] [PubMed]
85. Bagri, G.; Jones, G.V. The role of first person perspective and vivid imagery in memory for written narratives. *Educ. Psychol. Pr.* **2018**, *34*, 229–244. [CrossRef]
86. Libet, B. Unconscious cerebral initiative and the role of conscious will in voluntary action. *Behav. Brain Sci.* **1985**, *8*, 529. [CrossRef]
87. Bridgeman, B. Free will and the functions of consciousness. *Behav. Brain Sci.* **1985**, *8*, 540. [CrossRef]
88. Soon, C.S.; Brass, M.; Heinze, H.-J.; Haynes, J.-D. Unconscious determinants of free decisions in the human brain. *Nat. Neurosci.* **2008**, *11*, 543–545. [CrossRef] [PubMed]
89. Chen, Z.; Saunders, J.A. Volitional and automatic control of the hand when reaching to grasp objects. *J. Exp. Psychol. Hum. Percept. Perform.* **2018**, *44*, 953–972. [CrossRef]
90. Flanagan, O.; Polger, T. Zombies and the function of consciousness. *J. Conscious. Stud.* **1995**, *2*, 313–321.
91. Nelson, B.; Parnas, J.; Sass, L.A. Disturbance of Minimal Self (Ipseity) in Schizophrenia: Clarification and Current Status. *Schizophr. Bull.* **2014**, *40*, 479–482. [CrossRef]
92. Petersen, S.E.; Sporns, O. Brain Networks and Cognitive Architectures. *Neuron* **2015**, *88*, 207–219. [CrossRef]
93. Edlow, B.L.; McNab, J.A.; Witzel, T.; Kinney, H.C. The Structural Connectome of the Human Central Homeostatic Network. *Brain Connect.* **2016**, *6*, 187–200. [CrossRef] [PubMed]
94. Li, Q.; Wang, S.; Milot, E.; Bergeron, P.; Ferrucci, L.; Fried, L.P.; Cohen, A.A. Homeostatic dysregulation proceeds in parallel in multiple physiological systems. *Aging Cell* **2015**, *14*, 1103–1112. [CrossRef]
95. Andrews-Hanna, J.R.; Snyder, A.Z.; Vincent, J.L.; Lustig, C.; Head, D.; Raichle, M.E.; Buckner, R.L. Disruption of Large-Scale Brain Systems in Advanced Aging. *Neuron* **2007**, *56*, 924–935. [CrossRef] [PubMed]

Case Report

The Architect Who Lost the Ability to Imagine: The Cerebral Basis of Visual Imagery

Sandra Thorudottir [1,†], Heida M. Sigurdardottir [1,†], Grace E. Rice [2], Sheila J. Kerry [3], Ro J. Robotham [4], Alex P. Leff [3] and Randi Starrfelt [4,*]

1 Icelandic Vision Lab, Department of Psychology, University of Iceland, 102 Reykjavik, Iceland; sdt9@hi.is (S.T.); heidasi@hi.is (H.M.S.)
2 Cognition and Brain Sciences Unit, University of Cambridge, Cambridge CB27EF, UK; Grace.Rice@mrc-cbu.cam.ac.uk
3 Institute of Cognitive Neuroscience, University College London, London WC1N3AZ, UK; sheila.kerry10@googlemail.com (S.J.K.); a.leff@ucl.ac.uk (A.P.L.)
4 Department of Psychology, University of Copenhagen, 1726 Copenhagen, Denmark; jer@psy.ku.dk
* Correspondence: randi.starrfelt@psy.ku.dk
† These authors contributed equally to this work.

Received: 9 December 2019; Accepted: 19 January 2020; Published: 21 January 2020

Abstract: While the loss of mental imagery following brain lesions was first described more than a century ago, the key cerebral areas involved remain elusive. Here we report neuropsychological data from an architect (PL518) who lost his ability for visual imagery following a bilateral posterior cerebral artery (PCA) stroke. We compare his profile to three other patients with bilateral PCA stroke and another architect with a large PCA lesion confined to the right hemisphere. We also compare structural images of their lesions, aiming to delineate cerebral areas selectively lesioned in acquired aphantasia. When comparing the neuropsychological profile and structural magnetic resonance imaging (MRI) for the aphantasic architect PL518 to patients with either a comparable background (an architect) or bilateral PCA lesions, we find: (1) there is a large overlap of cognitive deficits between patients, with the very notable exception of aphantasia which only occurs in PL518, and (2) there is large overlap of the patients' lesions. The only areas of selective lesion in PL518 is a small patch in the left fusiform gyrus as well as part of the right lingual gyrus. We suggest that these areas, and perhaps in particular the region in the left fusiform gyrus, play an important role in the cerebral network involved in visual imagery.

Keywords: visual imagery; stroke; posterior cerebral artery; aphantasia; prosopagnosia; visual perception

1. Introduction

Thinking of a concept, whether it is a flower or a cat or even a unicorn, can bring up vivid, image-like experiences without external visual input. This is generally referred to as visual imagery or mental imagery, although the latter can extend to other senses (e.g., sound, smell, or touch). The basis of mental imagery has long been debated [1–3] and there is still uncertainty about its neural underpinnings.

Zeman and colleagues [4] gave the inability to generate mental imagery a name, aphantasia, and described individuals with congenital aphantasia who never had this ability. The loss of mental imagery following brain injury—acquired aphantasia—in individuals who had normal imagery before their injury is also well documented, dating back at least to Charcot and Bernard [5] (but see [6]). However, as noted by Farah [7], cases of acquired imagery deficits can be associated with a wide range of lesions (occipital, temporal, or parietal) in either hemisphere, and no other functional deficits consistently co-occurred with imagery loss with the exception of loss of (visual) dreaming. One plausible reason for this heterogeneity is that mental imagery is not a single phenomenon but can be divided into relatively distinct components, with different underlying anatomy. Some distinguish between a generation process, long-term visual memory, and an inspection process [7], or subsystems such as appearance-based (e.g., shape/color judgment) vs. spatial (e.g., mental navigation/scanning) imagery [8,9] (see also [10]). Supporting this, a meta-analysis of imaging studies showed that while several regions were coactivated during appearance-based and spatial imagery, the former mapped onto the ventral visual stream while the latter evoked specific activity in the dorsal stream [11].

It has been argued that the primary visual cortex (V1) plays a significant role in visual mental imagery [12,13]. Several studies have shown cortical activation in V1 during imagery tasks (e.g., [14–18]) and rTMS (repetitive transcranial magnetic stimulation) targeting V1 can disrupt visual imagery [15]. In addition, individual differences in mental imagery capability covary with differences in V1 surface area [19], V1 functional connectivity [20], and representational overlap between visual imagery and perception in the retinotopic cortex [21]. However, while patients with intact V1 can have severe impairments in mental imagery [22], seemingly intact imagery without a functioning V1 has also been reported [23,24] (see also [25]).

Thus, damage to V1 appears neither necessary nor sufficient for inducing imagery deficits. A review [26] of case studies suggested that extensive left temporal damage is necessary for a visual imagery deficit for object form or color (see also [11]), and more generally that high-level visual areas in the temporal lobe might be particularly important for visual imagery. The fact that patients have been reported to have both high level visual deficits and selective imagery loss in the same domain (e.g., severe problems in visual recognition and revisualization of faces, [27]), and that actual viewing and visual imagery for particular objects or object categories can evoke a similar pattern of activity in high-level ventral stream regions [21,28,29], is in alignment with the general idea of shared mechanisms between visual imagery and visual perception (for recent reviews, see [30,31]).

Visual imagery and perception however cannot share all mechanisms as there are patients on record with seemingly preserved mental imagery but impaired visual perception [32–36]. For example, case H.J.A. [32] suffered from visual agnosia, achromatopsia, prosopagnosia, alexia without agraphia and topographical impairments. Despite these deficits, H.J.A.'s mental imagery was relatively—albeit not completely—spared. The opposite pattern, impaired visual mental imagery but relatively normal visual perception, has also been reported [37,38]. An example is a patient who had suffered a left occipital and medial temporal infarct. While his visual recognition abilities were generally good, he showed apparent problems in mental imagery such as describing an elephant as having a "tiny waist" and having trouble with verifying sentences that required visual imagery (e.g., "A grapefruit is larger than an orange") [37].

Here we present patient PL518, an architect who reported almost complete loss of visual mental imagery following bilateral stroke in the areas supplied by the posterior cerebral artery (PCA). His responses on the Vividness of Visual Imagery Questionnaire (VVIQ, ad modum [39]) as well as a range of visuoperceptual tests are compared to three other patients with bilateral PCA stroke, as well as another architect with a large unilateral PCA stroke in the right hemisphere. We also compare the structural images of their lesions. The aim of the study is to: (a) describe the correspondence between the perceptual and neuropsychological profile of PL518 compared to the other patients, and (b) to delineate cerebral areas that are uniquely affected in the aphantasic patient and could thus play a fundamental role in the generation of visual imagery.

2. Materials and Methods

2.1. Participants

Patient PL518 and four other patients participated in this specific study. All were recruited as part of a larger study of PCA stroke (the Back of the Brain (BoB) project, described in [40]). 46 controls were included in the BoB project. All participants provided written, informed consent, and the project was approved by the ethical committees of Manchester (North West Research Ethics Committee; MREC 01/8/094) and UCL (London Queen Square Research Ethics Committee, UCL; 16/EM/0348). See Table 1 for demographics and background data for the included patients and controls. Additional background data as well as raw scores on the perceptual and neuropsychological tests can be found in Table S1.

Table 1. Demographic and lesion information for the five included patients and controls, and scores on basic tests. Handedness was measured by the short form of the Edinburgh Handedness Inventory (EHI) [41]; depression was measured with the short version of the Geriatric Depression Scale (GDS-15 [42]). General cognition was screened with the Oxford Cognitive Screen (OCS) [43], and the number of impaired subtests are listed. Digit span forward and backward was measured with the WAIS-IV UK [44] and total scores are listed. Basic motor reaction time (RT) was measured by responding to a bar of light presented horizontally on a screen (test described in [40]).

Participant	PL518	PL502	PL545	PM006	PM024	Controls Mean (SD)
Lesion Laterality	Bilat	Bilat	Bilat	Bilat	R	n/a
Age	52	55	62	67	66	62 (15)
Education (years)	18	17	16	12	16	15 (2)
Gender	M	M	M	F	M	24 F
Handedness (EHI)	−50	100	100	100	100	44 Right
Time Since Stroke (months)	35	20	14	36	10	n/a
Lesion Volume (cm^3)	52	23	57	24	112	n/a
Geriatric Depression Scale (GDS-15)	0	12	1	3	3	n/a
OCS-Impaired Subtests	0	1	6	1	1	n/a
Digit Span Forward (WAIS-IV max = 16)	10	13	15	11	11	11 (2)
Digit Span Backward (WAIS-IV max = 14)	6	6	8	8	6	8 (2)
Basic Motor RT (ms)	370	439	1195	665	686	398 (73)

PL518 suffered a bilateral PCA stroke 35 months before the current investigation. At that time, he had corrected to normal visual acuity and a slight visual field defect primarily affecting the parafovea of the upper left quadrant (see Table S1 for acuity and visual field data for all participants). He reported problems with seeing colors following his stroke and scored outside the normal range on a formal test of color perception (Farnsworth D-15 [45]). His intermediate vision (assessed with subtests from the L-POST [46]) was largely uncompromised, except for difficulties with figure-ground segmentation. His basic response time (RT) to visual stimuli (test described in [40]) was unaffected, and his auditory digit span forwards and backwards (WAIS-IV UK [44]) were within the normal range. PL518 reported severe problems in face recognition following his stroke and volunteered that he had problems recognizing his own face in the mirror. He also reported increased problems in finding his way around. Neuropsychological testing showed a clear deficit in face recognition affecting learning of new faces as well as judgment of familiarity and recognition of famous faces. In contrast, he performed within the normal range on several tests of object recognition, including perceptually challenging tests, with the notable exception of a memory test for houses (Cambridge House Memory Test, [47]) designed as an equivalent to the Cambridge Face Memory Test [48]. His word recognition accuracy was well within normal range, while response times in reading out loud and the effect of word length on RT was slightly but significantly elevated compared to controls.

In the context of the BoB project, PL518 spontaneously reported that his visual imagery was "gone" following his stroke, which led us to contact four additional patients either with similar lesions (bilateral PCA stroke) or a similar background (architect) for a follow-up interview about their visual imagery. The lesion location, size, and time since injury for these patients, as well as other background characteristics, are presented in Table 1 along with summary data from the control group.

2.2. Visual Imagery

As PL 518 was the focus of the study, a long and in-depth interview was also carried out about his visual imagery before and after his stroke. In order to get more information about his ability to store visual information in his mind, he was asked to carry out the Rey–Osterrieth Complex Figure Test [49] at the end of the interview. The other patients did not complete this test.

All five patients were asked to complete a version of the Vividness of Visual Imagery Questionnaire (VVIQ-modified, [39]), that is a modified version of the VVIQ [50]. VVIQ-modified has 16 items where participants are to imagine various scenarios (e.g., a relative or friend, a rising sun) and rate the vividness of their visual image on a five-point Likert scale where 1 indicates no image at all and 5 indicates that the image is perfectly clear and vivid. Scores on the VVIQ-modified can range from 16 to 80, with 16 representing the lowest possible imagery score and 80 the highest possible imagery score. Various versions of the VVIQ have been validated, and the questionnaire has been shown to be a valid psychometric tool for measuring the vividness of visual imagery with both high construct validity and internal consistency reliability [51–53]. PL518 and PM024 performed the questionnaire in the lab while the other three patients were interviewed over the telephone. The four control patients were also asked four general questions about their visual imagery, to compare with the interview of PL518. They were asked to answer the following questions with yes, no, or don't know: (1) can you imagine things visually in your mind? (if no: do you sometimes experience brief flashes of imagery?); (2) would you say that your memories have a visual aspect to them in your mind?; (3) do you see visual images in dreams?; and (4) has your visual imagery changed following your stroke? The normal controls did not perform the visual imagery questionnaire.

2.3. Neuropsychological and Experimental Tests

The BoB project is a comprehensive neuropsychological and imaging project investigating perceptual deficits following posterior brain injury [40]. A main aim of the overall project is to compare patient performance with faces, objects, and words. The main findings of the project are not yet published (paper in preparation, [54]). Here we report data from the five included patients and controls on tests and experiments selected to be comparable across categories for faces, objects, and words, and these are briefly described below. These experiments, as well as all other tests included in the project, are described in full in [40]. The experimental tests were run on laptop computers with screen resolution of 1366 × 768, or on desktop computers with a screen resolution of 1920 × 1080.

2.3.1. Delayed Matching and Surprise Recognition of Words, Objects and Faces—The WOF Test

This novel paradigm is designed to test immediate and delayed memory for words, objects, and faces (WOF). In the first part (delayed matching), participants were asked to decide whether two sequentially presented images varying in size are the same or not. There were 48 trials for each stimulus type and both accuracy and RTs are measured. The second part (surprise recognition) followed after a short break (where participants performed an unrelated task, the Farnsworth D-15). Here, participants were asked to decide whether they saw the presented stimuli in the delayed matching task or not. There were 12 trials, and accuracy and RTs were measured. In total, 12 measures were derived from this task: 2 metrics (accuracy and RT) * 3 stimulus types (words, objects, faces) * 2 paradigms (delayed matching and surprise recognition). See [40] for a more detailed description.

2.3.2. Familiarity Decisions

Familiarity decision tasks were run for faces, objects, and words. For faces, participants were asked to decide whether a presented face was famous or not (80 trials in total). For objects, we used a 72-trial version of a well-studied object decision task [55], presenting line drawings of real objects and chimeric non-objects. Participants were asked to decide if the picture represents a real object or a non-object. For words, we used a lexical decision task with 60 trials, where participants were asked to decide whether the presented letter string represented a real word or a pseudoword. For all three familiarity decision tests, both accuracy and RTs for correctly categorized familiar items (famous faces, real objects, real words) were analyzed.

2.3.3. Naming of Familiar Items

Tests of picture naming (line drawings), face naming (famous faces) and word reading (regular words). For pictures and words, both accuracy and RTs for correctly named items are analyzed (a voice key was used for RT measurement). For famous faces, only accuracy is recorded as measuring RTs in face naming tasks is complicated by participants making other verbal responses than names (e.g., "it's that guy from the Parliament ... ").

2.4. Structural MRI: Lesions

Structural brain imaging data were acquired from all subjects. Structural scans were acquired on two 3T Phillips Achieva scanners with 32-channel head-coils and a SENSE factor of 2.5 in London and Manchester. A high-resolution T1 weighted structural scan was acquired for spatial normalization, including 260 slices covering the whole brain with TR = 8.4 ms, TE = 3.9 ms, flip angle = 8 degrees, FOV = 240×191 mm^2, resolution matrix = 256×206 and voxels size = $0.9 \times 1.7 \times 0.9$ mm^3. Automated outlines of the area affected by stroke were generated using Seghier et al.'s modified segmentation–normalization procedure [56]. Segmented images were smoothed with an 8mm full-width half maximum Gaussian kernel and submitted to the automated routines for lesion identification and definition modules using the default parameters. The automated method involves initial segmentation and normalizing into tissue classes of grey matter, white matter, cerebro-spinal fluid (CSF), and an extra tissue class for the presence of a lesion. After smoothing, voxels that emerge as outliers relative to the normal population are identified and the union of these outliers provides the "fuzzy lesion map", from which the lesion outline is derived. The generated images were used to create the lesion overlap maps.

3. Results

3.1. Visual Imagery

In the clinical interview, PL518 reported an almost complete absence of visual imagery following his stroke. This was in stark contrast to his (in his own opinion) above average ability for visual imagery before his stroke that he had relied upon in his work as an architect. He said: "Before, my visualization abilities were pretty impressive. At my work, I could visualize and remember things that most people had not thought about. I would be sitting there and I would say, well, you can't do X, Y and Z, because you've got this happening here and there. Now I have to look at the drawing and work my way through it." During the interview, he also described how it had felt to do a mental rotation task: "I cannot do it as quickly or the same way as I would have done before my stroke. Before, bang, I would just know the answer. Now it is a much more conscious process. It's almost as though I physically am trying to move things inside my head." He was then asked whether his difficulty with mental rotation affected his ability to work as an architect, to which he responded: "Well I just do everything on the computer. That is one of the advantages of us using computers for these sorts of thing nowadays. You can see the stuff happen." He also described how he is just about able to imagine very simple shapes, but this is done using something akin to motor or spatial imagery and he struggles to imagine more than one shape at a time: "If I tried to visualize shapes like a square, pyramid or sphere lined up next to each other, and I try and focus with a kind of spotlight on the corner of one shape, I can mentally trace a line around the shape. But as soon as I focus on one shape, the others disappear." When asked if he could imagine an elephant, he seemed to mostly think of the abstract concept of an elephant: "I can think of elephants, iconic elephants like Babar or Elmer, but I can only visualize bits of them. It's almost painful." When asked to describe the place he stayed during his last holiday and its surroundings he provided few very vague details about a couple of the bars from the street they had lived on, and he apparently did not visually imagine himself there: "I am recalling almost like a list. I do the same when going somewhere. I have to remember a list".

PL518's copy and retention of the Rey figure are shown in Figure 1. The drawings were scored for accuracy according to the Taylor's (1969) method described in Spreen and Straus (1991). 35/36 points were given for the copy and 18/36 for the three-minute recall. While these scores are within the normal range, one could have expected patient PL518, with his background as an architect, to have adopted a more structured approach to drawing the figure in the recall condition. This drawing not only lacks many details but also some of the core elements. Also, some of the included elements are placed incorrectly.

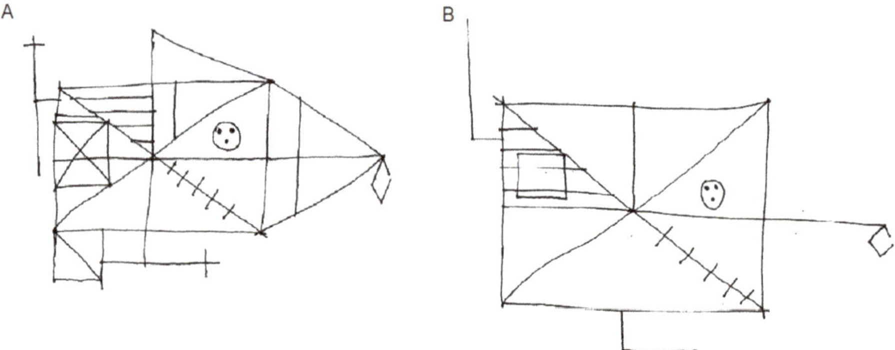

Figure 1. PL518's performance on the Rey Complex Figure Test. (**A**): Copy. (**B**): three-minute recall.

PL518's complaints regarding his visual imagery were also clearly reflected in his responses on the VVIQ-modified where he scored 18 (i.e., a mean score of 1.13 per item), corresponding to minimal imagery [39]. None of the other patients reported any changes in the nature or vividness of their visual imagery following their strokes–neither in the VVIQ-modified nor the general questions; they all responded yes to the first three general questions about being able to see images in their minds, and no when asked if their visual imagery had changed following their stroke. Their respective scores on the VVIQ-modified were: PL502: 49 (mean: 3.06); PL545: 53 (mean: 3.31); PM006: 72 (mean: 4.5); PM024: 76 (mean: 4.75). See Appendix A Table A1 for the patients' responses to the individual questions.

3.2. Neuropsychological and Experimental Tests

For the accuracy measures, PL518 is clearly impaired with faces, and shows a deficit (performing more than two standard deviations (SDs) from the control mean) on most individual face measures. He performs within the low–normal range on the object tests and is clearly on level with controls in the tests with word stimuli. For the RT measures, PL518 shows a deficit on most face measures (note that his RTs in the surprise recognition test may not be a good indicator of severity, as his accuracy in this test was very low). He responds with latencies within the normal range on the object tests but shows elevated RTs in the lexical decision and word reading tests.

Comparing the neuropsychological profile of PL518 to the other included patients, we find that one or more of them show deficits on the same tests/measures and in the cognitive domain(s) as PL518 (see Figure 2 for an illustration of their cognitive profiles on the selected tests, and Table S1 for an overview of test results). A comparison of the neuropsychological profile of the two architects (PL518 and PM024) shows that PM024 (with no aphantasia) shows the same pattern of performance as PL518 on most tests, including measures of face recognition, object recognition and word reading. Indeed, there is no measure on which PL518 shows a clear deficit, where PM024 is clearly within the normal range (see Figure 2). The key difference between the two patients, then, is in the measure of their visual imagery. Comparing PL518 to the three other bilateral patients (Figure 2 and Figure S1), again there is no domain where PL518 is clearly impaired where the other patients are consistently within the normal range. In comparison to the three bilateral patients too, then, the key difference is in visual imagery.

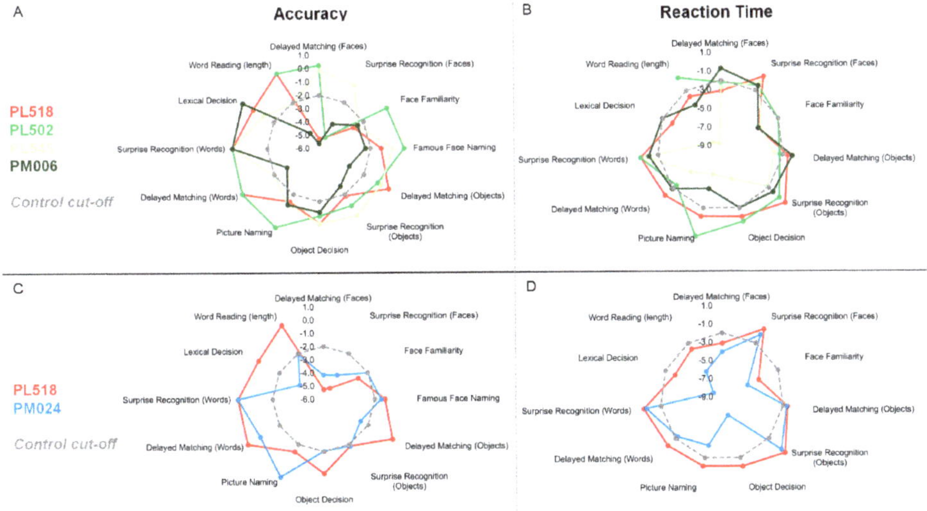

Figure 2. Radar plots showing the results of PL518 (in red) and the other patients on all the included measures of object, word, and face recognition. Numbers denote z-scores based on the control means and SDs for the respective tests. Impaired performance (>−2 SDs from the control mean) is marked by the dotted grey line, and scores closer to the center are more impaired (represents lower accuracy and slower RTs). Left panel (**A**,**C**) shows accuracy, right panel (**B**,**D**) shows RTs. Upper panel (**A**,**B**) shows PL518 vs. PM024 (architect with right hemisphere lesion). Lower panel (**C**,**D**) shows PL518 vs. the other bilateral patients. See individual radar plots comparing PL518 individually to bilateral patients in Figure S1.

3.3. Lesion Localisation

PL518's lesion is most extensive on the right side, including damage to the occipital pole, the lingual gyrus, the whole fusiform gyrus and extending anteriorly to the parahippocampal region. On the left side, the lesion affects only the medial fusiform gyrus and lingual gyrus, while the left occipital pole, and lateral portions of the fusiform gyrus are spared. See Figure 3 and Table 2 for comparisons of lesion localization for PL518 and the other patients.

First, comparing the lesions of PL518 to the architect without aphantasia (PM024) shows that PL518 has selective left hemisphere posterior medial fusiform damage extending medially and anteriorly along the collateral sulcus, and selective right hemisphere damage to the superior medial lingual gyrus. Second, comparing the lesion of PL518 to the three patients with bilateral strokes but no aphantasia shows that PL518 has selective damage in the right fusiform gyrus and a portion of the right lingual gyrus, and additional smaller areas of selective damage in PL518 are found in the left fusiform gyrus. Combined, these comparisons reveal only small areas of selective damage in PL518 in the right lingual gyrus and left posterior medial fusiform gyrus.

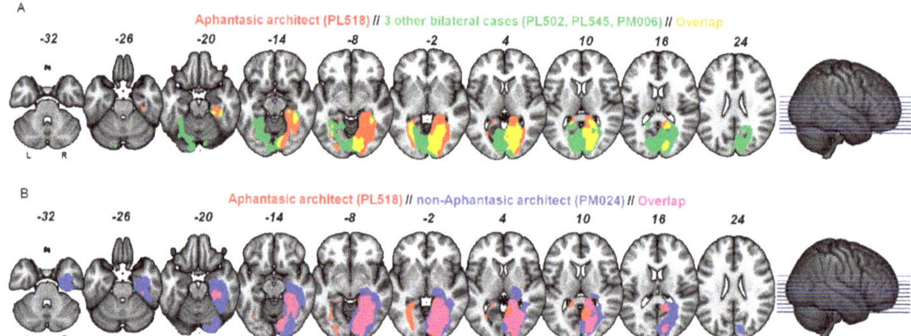

Figure 3. (**A**) shows lesion overlap between PL518 and the other bilateral patients. (**B**) shows lesion overlap between PL518 and PM024. Left hemisphere depicted on the left. The lesions are in MNI space overlaid on the MNI152 template. See Figure S2 for individual comparisons of PL518 and the bilateral patients.

Table 2. Comparison of regions of interest within the occipital and temporal lobes affected in PL518 compared to other patients. The fusiform gyrus (FG) was segmented into four regions (FG1-4: corresponding to posterior medial, posterior lateral, anterior medial and anterior lateral, respectively) according to Lorenz and colleagues [57]. The occipital pole and the lingual gyrus were defined using a conventional atlas [58], and the parahippocampal region was identified using the images from Bouyeure and colleagues [59]. An x indicates that at least 10% of the corresponding region of interest was affected by a patient's stroke.

	Patient	**PL518**	**PL502**	**PL545**	**PM006**	**PM024**
	Laterality	**Bilateral**	**Bilateral**	**Bilateral**	**Bilateral**	**Right**
Left hemisphere	Occipital Pole			x	x	
	FG 1	x	x			
	FG 2		x			
	FG 3	x	x			
	FG 4					
	Lingual Gyrus	x	x	x	x	
	Parahipp. Gyrus		x			
Right hemisphere	Occipital Pole	x		x	x	x
	FG 1	x				x
	FG 2	x				x
	FG 3	x	x			x
	FG 4	x	x			x
	Lingual Gyrus	x		x	x	x
	Parahipp. Gyrus	x	x			x

4. Discussion

The present study reports case PL518, an architect who lost his ability for visual imagery following a bilateral PCA stroke 35 months prior to this investigation. We compare his performance across a range of perceptual and cognitive tests and a visual imagery questionnaire with four other PCA stroke patients, an architect with a large right hemisphere lesion and three bilateral cases. PL518's profile on the perceptual and cognitive tests was similar to other cases with the exception that PL518 reported severe visual imagery problems following his stroke. Lesion profiles were also comparable with the exception that PL518 showed selective damage in the right lingual gyrus and left medial posterior fusiform gyrus. It is tempting to suggest that these are both candidate regions for specific involvement in visual imagery.

However, Bogousslavsky and colleagues [60] described a man whose lingual gyrus was destroyed in both hemispheres, while only the middle third of the fusiform gyrus on the left side was affected. His visual imagery was intact for colors, faces (human and animal) and places (streets). The authors concluded that the fusiform gyrus and underlying white matter, rather than the lingual gyrus, was a principal structure for color integration, face recognition, visuo-verbal processing, and corresponding visual imagery. The fact that the current primary case, PL518, had selective damage to the left fusiform gyrus is also more in alignment with other research indicating that left hemisphere regions are more consistently implicated in generating mental imagery than corresponding right hemisphere regions [4,7,22,24,61–68].

A seeming counterexample comes from de Gelder and colleagues [69]. They described patient TN who had bilateral cortical blindness due to lesions in the primary visual cortices in both hemispheres. The lesion also reached some high-level visual ventral areas, including parts of the left posterior fusiform gyrus. Despite this damage, de Gelder and colleagues [69] argued that TN was able to generate visual mental imagery. However, judging from the lesion reconstruction (their Figure 2), the left medial posterior fusiform might have been at least partially spared in this patient. Also, the imagery tasks used involved a significant motor or action component, and correspondingly TN's functional activation pattern in the imagery conditions was primarily fronto-parietal.

Fitting with a role of the left fusiform gyrus in visual imagery, some developmental prosopagnosics appear to have functional abnormalities in this region [70–72] as well as reduced or absent mental imagery, not only for faces but also for objects and scenes [73]. Barton and Cherkasova [74] examined face imagery in prosopagnosics for featural imagery (questions regarding facial features, e.g., "Who has a wider mouth: Sophia Loren or Ingrid Bergman?") as well as facial configurations (questions on overall face shape or configuration, e.g., "Who has the more angular face: George Washington or Abraham Lincoln?"). In acquired prosopagnosics, they found that right-sided occipito-temporal lesions affected imagery for facial configuration but not for facial features, while bilateral lesions additionally impaired imagery for facial features [74]. This fits well with the left fusiform gyrus responding more to facial features while the right fusiform gyrus is more involved in configural processing [75]. It is possible that the generation of mental imagery heavily relies on the assembly of separately stored visual features or parts, and that this generation of multipart images specifically taxes left hemisphere regions [37,66]. This is consistent with PL518's description of the fragmented minimal visual imagery that he possibly still has (e.g., visualizing bits of elephants).

Compared to before his stroke, PL518 seems to make greater use of verbal strategies (e.g., recalling a list). If PL518 still has some mental imagery, it nonetheless mostly seems to be based on an altered strategy which could be described as motor, action-based, or spatial, such as mentally tracing a line around a shape or doing mental rotation by physically trying to move things inside the head. This is reminiscent of patient MX [4] who also reported the loss of the experience of visual imagery as well as an unusual or altered strategy when attempting a mental rotation task, where he needed to match individual blocks and angles perceptually when making his decision.

The two architects, PL518 and PM024, had similar functional deficits, including prosopagnosia, but described vastly different visual imagery (minimal vs. very clear and lively). It is tempting to speculate, therefore, that the additional left hemisphere affection in PL518 contributes significantly to his disruption of imagery. In particular, the small patch in the medial left fusiform gyrus where PL518 has unique damage compared to all the four other patients presents as a good candidate for playing a critical part in the generation of visual mental imagery. While our findings indeed suggest that this region is an important node in the cerebral network underlying visual imagery, other areas, including right hemisphere ventral occipito-temporal areas, left hemisphere areas further anterior in the temporal lobe (see e.g., [74]), more posterior areas in the left occipital lobe, and regions outside of the ventral visual stream are also likely to partake in at least some aspects of visual imagery. For example, while mental imagery generation might mainly depend on structures in the posterior left hemisphere, right parietal regions have been found to be important for spatial comparisons of the contents of visual imagery [76], see also [77]. The right hemisphere could also have some ability to generate visual imagery for overall shape [66], and had we included sensitive measures of configural processing deficits in mental imagery in addition to the VVIQ, it is possible that subtle deficits in PM024 could have been discovered. It is also worth noting that the aphantasic architect PL518 had bilateral damage, while mental imagery generation could possibly be taken over by the right hemisphere in cases of unilateral left hemisphere disruption [76].

The most commonly used questionnaire to measure mental imagery is various versions of the VVIQ [39,50]. The VVIQ has good psychometric qualities and vividness correlates with some other behavioral and neural measures of visual imagery [53,78,79]. These questionnaires do have their limitations, though, as they rely on self-reporting and only measure overall vividness of visual mental imagery. Mental imagery is, however, of a multimodal nature [80] and includes for example smell, touch, sound and taste. Also, there are several different aspects of visual imagery, and in order to capture this more completely, a measure would need to include items specifically for spatial imagery, as well as imagery for colors, objects, places, faces, and even subsets of these such as featural vs. configural face imagery. More fine-grained mental imagery questionnaires and additional behavioral measures that likely rely on mental imagery, such as the clock task [81–83], the taller/wider task [66,83], or mental letter construction [84], animal tails test [8], drawing objects from memory [85,86] and the binocular–rivalry technique [87,88], would provide further insights into whether mental imagery deficits are due to a loss of all imagery across modalities, specific loss of visual imagery, or specific loss of subcomponents of visual imagery. Such specific aspects of mental imagery were not directly assessed in the present study.

It is still debated whether imagery and perception may be dissociated, or whether they depend on common networks. In one sense, the current results support the former as some patients with heavy damage to ventral stream areas and associated problems with visual cognition nonetheless appear to have intact visual imagery. Our neuropsychological approach suggests that some ventral stream regions might not be necessary for visual imagery despite containing information on imagined objects [21,28,29,89,90]. On the other hand, the areas specifically associated with PL518's visual imagery loss are better known for their role in visual perception. A key difference between imagery and perception could however lie in their different network dynamics where imagery is dominated by top-down feedback [21,89,90]; this could even map onto different cortical layers within the same region [91,92]. Even if a region serves both perception and imagery, is it still possible that distinct computations and separable subpopulations of neurons are involved.

It should finally be noted that individual differences in premorbid ability for imagery might play a role in the effects of stroke on these abilities. PL518 reported that his abilities for visual mental imagery had been above average before his stroke. These abilities had enabled him to visualize the spatial and visual attributes of buildings and rooms in rich details and contributed greatly to his achievements as an architect. This fits a general pattern noted by Farah [7] where many cases of acquired deficits in visual imagery involved people whose day-to-day activities had likely demanded visualization. As the normal variability in visual imagery from congenital aphantasia to hyperphantasia becomes better understood, this factor may perhaps help explain variability in the effect of brain injury on visual imagery.

5. Conclusions

While several brain regions in both hemispheres are involved in different aspects of mental imagery, our results indicate that the right lingual gyrus and especially the left posterior medial fusiform gyrus are candidate regions for specific involvement in *visual* imagery. These regions were only affected in the aphantasic architect PL518 compared to non-aphantasic patients with comparable cognitive and perceptual deficits.

Supplementary Materials: The following are available online at http://www.mdpi.com/2076-3425/10/2/59/s1, Table S1. Individual neuropsychological patient raw data. Figure S1. Radar plots comparing individual bilateral patients' neuropsychological profile to PL518. Figure S2. Individual comparison of bilateral patients' structural MRI with PL518.

Author Contributions: Conceptualization, S.J.K., G.E.R., A.P.L., R.J.R., and R.S.; methodology, S.J.K., G.E.R., A.P.L., R.J.R., and R.S.; software, R.J.R., R.S., S.J.K., and G.E.R.; formal analysis, S.J.K., G.E.R., and R.S.; investigation, G.E.R., S.J.K., R.J.R., and R.S.; data curation, S.J.K., R.S., R.J.R., and G.E.R.; writing—original draft preparation, S.T. and H.M.S.; writing—review and editing, S.T., H.M.S., R.S., G.E.R., R.J.R., A.P.L., and S.J.K.; visualization, S.J.K., G.E.R.; supervision, R.S:; project administration, R.S.; funding acquisition, R.S. All authors have read and agreed to the published version of the manuscript.

Funding: This work was funded by a grant from Independent Research Fund Denmark (Sapere Aude) to R.S. (Grant no. DFF–4180-00201).

Acknowledgments: We wish to thank Matthew A. Lambon Ralph for contributing to the BoB-project and thus making the present work possible, and Fakutsi for support during the preparation of this manuscript.

Conflicts of Interest: The authors declare no conflict of interest. The funders had no role in the design of the study; in the collection, analyses, or interpretation of data; in the writing of the manuscript, or in the decision to publish the results.

Appendix A

Table A1. Individual responses on the Vividness of Imagery Questionnaire (VVIQ)-modified [39]. Total and mean scores in **bold.**

VVIQ Question	PL518	PL502	PL545	PM006	PM024
1	2	4	4	5	5
2	1	3	3	4	5
3	2	1	2	4	4
4	1	1	4	5	5
5	1	4	4	5	5
6	1	3	4	4	5
7	1	3	3	5	5
8	1	5	4	5	5
9	1	3	2	4	4
10	1	1	2	4	5
11	1	3	2	4	4
12	1	5	3	4	5
13	1	4	4	5	5
14	1	3	4	5	4
15	1	2	4	5	5
16	1	4	4	4	5
Total	**18**	**49**	**53**	**72**	**76**
Mean	**1.13**	**3.06**	**3.31**	**4.50**	**4.75**

References

1. Pylyshyn, Z.W. What the mind's eye tells the mind's brain: A critique of mental imagery. *Psychol. Bull.* **1973**, *80*, 1–24. [CrossRef]
2. Pylyshyn, Z.W. The imagery debate: Analogue media versus tacit knowledge. *Psychol. Rev.* **1981**, *88*, 16–45. [CrossRef]
3. Kosslyn, S.M. The medium and the message in mental imagery: A theory. *Psychol. Rev.* **1981**, *88*, 46–66. [CrossRef]
4. Zeman, A.Z.; Della Sala, S.; Torrens, L.A.; Gountouna, V.E.; McGonigle, D.J.; Logie, R.H. Loss of imagery phenomenology with intact visuo-spatial task performance: A case of 'blind imagination'. *Neuropsychologia* **2010**, *48*, 145–155. [CrossRef]
5. Charcot, J.M.; Bernard, D. Un cas de suppression brusque et isolée de la vision mentale des signes et des objets (formes et couleurs). *Le Progrés Médical* **1883**, *11*, 568–571.

6. Zago, S.; Allegri, N.; Cristoffanini, M.; Ferrucci, R.; Porta, M.; Priori, A. Is the Charcot and Bernard case (1883) of loss of visual imagery really based on neurological impairment? *Cogn. Neuropsychiatry* **2011**, *16*, 481–504. [CrossRef]
7. Farah, M.J. The neurological basis of mental imagery: A componential analysis. *Cognition* **1984**, *18*, 245–272. [CrossRef]
8. Farah, M.J.; Hammond, K.M.; Levine, D.N.; Calvanio, R. Visual and spatial mental imagery: Dissociable systems of representation. *Cogn. Psychol.* **1988**, *20*, 439–462. [CrossRef]
9. Levine, D.N.; Warach, J.; Farah, M. Two visual systems in mental imagery: Dissociation of "what" and "where" in imagery disorders due to bilateral posterior cerebral lesions. *Neurology* **1985**, *35*, 1010. [CrossRef]
10. Kosslyn, S.M. Seeing and imagining in the cerebral hemispheres: A computational approach. *Psychol. Rev.* **1987**, *94*, 148–175. [CrossRef]
11. Mazard, A.; Tzourio-Mazoyer, N.; Crivello, F.; Mazoyer, B.; Mellet, E. A PET meta-analysis of object and spatial mental imagery. *Eur. J. Cogn. Psychol.* **2004**, *16*, 673–695. [CrossRef]
12. Kosslyn, S.M.; Ganis, G.; Thompson, W.L. Neural foundations of imagery. *Nat. Rev. Neurosci.* **2001**, *2*, 635–642. [CrossRef] [PubMed]
13. Kosslyn, S.M.; Thompson, W.L. When is early visual cortex activated during visual mental imagery? *Psychol. Bull.* **2003**, *129*, 723–746. [CrossRef] [PubMed]
14. Chen, W.; Kato, T.; Zhu, X.H.; Ogawa, S.; Tank, D.W.; Ugurbil, K. Human primary visual cortex and lateral geniculate nucleus activation during visual imagery. *Neuroreport* **1998**, *9*, 3669–3674. [CrossRef]
15. Kosslyn, S.M.; Pascual-Leone, A.; Felician, O.; Camposano, S.; Keenan, J.P.; Ganis, G.; Sukel, K.E.; Alpert, N.M. The role of area 17 in visual imagery: Convergent evidence from PET and rTMS. *Science* **1999**, *284*, 167–170. [CrossRef]
16. Klein, I.; Paradis, A.L.; Poline, J.B.; Kosslyn, S.M.; Le Bihan, D. Transient activity in the human calcarine cortex during visual-mental imagery: An event-related fMRI study. *J. Cogn. Neurosci.* **2000**, *12*, 15–23. [CrossRef]
17. Ganis, G.; Thompson, W.L.; Kosslyn, S.M. Brain areas underlying visual mental imagery and visual perception: An fMRI study. *Cogn. Brain Res.* **2004**, *20*, 226–241. [CrossRef]
18. Slotnick, S.D.; Thompson, W.L.; Kosslyn, S.M. Visual mental imagery induces retinotopically organized activation of early visual areas. *Cereb. Cortex* **2005**, *15*, 1570–1583. [CrossRef]
19. Bergmann, J.; Genç, E.; Kohler, A.; Singer, W.; Pearson, J. Smaller primary visual cortex is associated with stronger, but less precise mental imagery. *Cereb. Cortex* **2016**, *26*, 3838–3850. [CrossRef]
20. Dijkstra, N.; Zeidman, P.; Ondobaka, S.; van Gerven, M.A.; Friston, K. Distinct top-down and bottom-up brain connectivity during visual perception and imagery. *Sci. Rep.* **2017**, *7*, 5677. [CrossRef]
21. Lee, S.H.; Kravitz, D.J.; Baker, C.I. Disentangling visual imagery and perception of real-world objects. *Neuroimage* **2012**, *59*, 4064–4073. [CrossRef] [PubMed]
22. Moro, V.; Berlucchi, G.; Lerch, J.; Tomaiuolo, F.; Aglioti, S.M. Selective deficit of mental visual imagery with intact primary visual cortex and visual perception. *Cortex* **2008**, *44*, 109–118. [CrossRef] [PubMed]
23. Bridge, H.; Hicks, S.L.; Xie, J.; Okell, T.W.; Mannan, S.; Alexander, I.; Cowey, A.; Kennard, C. Visual activation of extra-striate cortex in the absence of V1 activation. *Neuropsychologia* **2010**, *48*, 4148–4154. [CrossRef] [PubMed]
24. Bridge, H.; Harrold, S.; Holmes, E.A.; Stokes, M.; Kennard, C. Vivid visual mental imagery in the absence of the primary visual cortex. *J. Neurol.* **2012**, *259*, 1062–1070. [CrossRef] [PubMed]
25. Chatterjee, A.; Southwood, M.H. Cortical blindness and visual imagery. *Neurology* **1995**, *45*, 2189–2195. [CrossRef]
26. Bartolomeo, P. The relationship between visual perception and visual mental imagery: A reappraisal of the neuropsychological evidence. *Cortex* **2002**, *38*, 357–378. [CrossRef]
27. Shuttleworth, E.C., Jr.; Syring, V.; Allen, N. Further observations on the nature of prosopagnosia. *Brain Cogn.* **1982**, *1*, 307–322. [CrossRef]
28. O'Craven, K.M.; Kanwisher, N. Mental imagery of faces and places activates corresponding stimulus-specific brain regions. *J. Cogn. Neurosci.* **2000**, *12*, 1013–1023. [CrossRef]
29. Reddy, L.; Tsuchiya, N.; Serre, T. Reading the mind's eye: Decoding category information during mental imagery. *Neuroimage* **2010**, *50*, 818–825. [CrossRef]

30. Dijkstra, N.; Bosch, S.E.; van Gerven, M.A. Shared neural mechanisms of visual perception and imagery. *Trends Cogn. Sci.* **2019**, *23*, 423–434. [CrossRef]
31. Pearson, J.; Kosslyn, S.M. The heterogeneity of mental representation: Ending the imagery debate. *Proc. Natl. Acad. Sci. USA* **2015**, *112*, 10089–10092. [CrossRef]
32. Riddoch, M.J.; Humphreys, G.W. A case of integrative visual agnosia. *Brain* **1987**, *110*, 1431–1462. [CrossRef]
33. Humphreys, G.W.; Riddoch, M.J. Routes to object constancy: Implications from neurological impairments of object constancy. *Q. J. Exp. Psychol.* **1984**, *36*, 385–415. [CrossRef] [PubMed]
34. Riddoch, M.J.; Humphreys, G.W.; Gannon, T.; Blott, W.; Jones, V. Memories are made of this: The effects of time on stored visual knowledge in a case of visual agnosia. *Brain* **1999**, *122*, 537–559. [CrossRef] [PubMed]
35. Behrmann, M.; Winocur, G.; Moscovitch, M. Dissociation between mental imagery and object recognition in a brain-damaged patient. *Nature* **1992**, *359*, 636–637. [CrossRef] [PubMed]
36. Behrmann, M.; Moscovitch, M.; Winocur, G. Intact visual imagery and impaired visual perception in a patient with visual agnosia. *J. Exp. Psychol. Hum. Percept. Perform.* **1994**, *20*, 1068–1087. [CrossRef]
37. Farah, M.J.; Levine, D.N.; Calvanio, R. A case study of mental imagery deficit. *Brain Cogn.* **1988**, *8*, 147–164. [CrossRef]
38. Goldenberg, G. Loss of visual imagery and loss of visual knowledge—A case study. *Neuropsychologia* **1992**, *30*, 1081–1099. [CrossRef]
39. Zeman, A.Z.; Dewar, M.; Della Sala, S. Lives without imagery—Congenital aphantasia. *Cortex* **2015**, *73*, 378–380. [CrossRef]
40. Robotham, R.J. The Neuropsychology of Stroke in the Back of the Brain: Clinical and Cognitive Aspects. Ph.D. Dissertation, University of Copenhagen, Faculty of Social Science, Department of Psychology, Copenhagen, Denmark, 2018.
41. Veale, J.F. Edinburgh handedness inventory–short form: A revised version based on confirmatory factor analysis. Laterality: Asymmetries of Body. *Brain Cogn.* **2014**, *19*, 164–177.
42. Yesavage, J.A.; Sheikh, J.I. Geriatric Depression Scale (GDS): Recent evidence and development of a shorter version. *Clin. Gerontol.* **1986**, *5*, 165–173. [CrossRef]
43. Demeyere, N.; Riddoch, M.J.; Slavkova, E.D.; Bickerton, W.L.; Humphreys, G.W. The Oxford Cognitive Screen (OCS): Validation of a stroke-specific short cognitive screening tool. *Psychol. Assess.* **2015**, *27*, 883. [CrossRef] [PubMed]
44. Wechsler, D. *Wechsler Adult Intelligence Scale Fourth UK Edition*; The Psychology Corporation: London, UK, 2010.
45. Linksz, A. The Farnsworth panel D-15 test. *Am. J. Ophthalmol.* **1966**, *62*, 27–37. [CrossRef]
46. Torfs, K.; Vancleef, K.; Lafosse, C.; Wagemans, J.; de-Wit, L. The Leuven Perceptual Organization Screening Test. (L-POST), an online test to assess mid-level visual perception. *Behav. Res. Methods* **2014**, *46*, 472–487. [CrossRef] [PubMed]
47. Martinaud, O.; Pouliquen, D.; Gérardin, E.; Loubeyre, M.; Hirsbein, D.; Hannequin, D.; Cohen, L. Visual agnosia and posterior cerebral artery infarcts: An anatomical-clinical study. *PLoS ONE* **2012**, *7*, e30433. [CrossRef]
48. Duchaine, B.; Nakayama, K. The Cambridge Face Memory Test: Results for neurologically intact individuals and an investigation of its validity using inverted face stimuli and prosopagnosic participants. *Neuropsychologia* **2006**, *44*, 576–585. [CrossRef] [PubMed]
49. Rey, A. L'examen psychologique dans les cas d'encéphalopathie traumatique.(Les problems.). *Arch. Psychol.* **1941**, *28*, 215–285.
50. Marks, D.F. Visual imagery differences in the recall of pictures. *Br. J. Psychol.* **1973**, *64*, 17–24. [CrossRef]
51. Campos, A.; Pérez-Fabello, M.J. Psychometric quality of a revised version Vividness of Visual Imagery Questionnaire. *Percept. Mot. Ski.* **2009**, *108*, 798–802. [CrossRef]
52. Campos, A. Internal consistency and construct validity of two versions of the Revised Vividness of Visual Imagery Questionnaire. *Percept. Mot. Ski.* **2011**, *113*, 454–460. [CrossRef]
53. Morrison, R.G.; Wallace, B. Imagery vividness, creativity and the visual arts. *J. Ment. Imag. N. Y. Int. Imag. Assoc.* **2001**, *25*, 135–152.
54. Rice, G.E.; Keryy, S.J.; Robotham, R.J.; Leff, A.P.; Lambon Ralph, M.A.; Starrfelt, R. Revealing the spectrum of visual perceptual function following posterior cerebral artery stroke. In preparation.

55. Gerlach, C. Normal and abnormal category-effects in visual object recognition: A legacy of Glyn W. Humphreys. *Vis. Cogn.* **2017**, *25*, 60–78. [CrossRef]
56. Seghier, M.L.; Ramlackhansingh, A.; Crinion, J.; Leff, A.P.; Price, C.J. Lesion identification using unified segmentation-normalisation models and fuzzy clustering. *NeuroImage* **2008**, *41*, 1253–1266. [CrossRef] [PubMed]
57. Lorenz, S.; Weiner, K.S.; Caspers, J.; Mohlberg, H.; Schleicher, A.; Bludau, S.; Eickhoff, S.B.; Grill-Spector, K.; Zilles, K.; Amunts, K. Two new cytoarchitectonic areas on the human mid-fusiform gyrus. *Cereb. Cortex* **2015**, *27*, 373–385. [CrossRef]
58. Duvernoy, H.M. *The Human Brain: Surface, Three-Dimensional Sectional Anatomy with MRI, and Blood Supply*; Springer: Vienna, Austria, 1999.
59. Bouyeure, A.; Germanaud, D.; Bekha, D.; Delattre, V.; Lefèvre, J.; Pinabiaux, C.; Mangin, J.F.; Rivière, D.; Fischer, C.; Chiron, C.; et al. Three-dimensional probabilistic maps of mesial temporal lobe structures in children and adolescents' brains. *Front. Neuroanat.* **2018**, *12*, 98. [CrossRef]
60. Bogousslavsky, J.; Miklossy, J.; Deruaz, J.P.; Assal, G.; Regli, F. Lingual and fusiform gyri in visual processing: A clinico-pathologic study of superior altitudinal hemianopia. *J. Neurol. Neurosurg. Psychiatry* **1987**, *50*, 607–614. [CrossRef]
61. Bartolomeo, P. The neural correlates of visual mental imagery: An ongoing debate. *Cortex* **2008**, *44*, 107–108. [CrossRef]
62. D'Esposito, M.; Detre, J.A.; Aguirre, G.K.; Stallcup, M.; Alsop, D.C.; Tippet, L.J.; Farah, M.J. A functional MRI study of mental image generation. *Neuropsychologia* **1997**, *35*, 725–730. [CrossRef]
63. Farah, M.J. The laterality of mental image generation: A test with normal subjects. *Neuropsychologia* **1986**, *24*, 541–551. [CrossRef]
64. Farah, M.J.; Gazzaniga, M.S.; Holtzman, J.D.; Kosslyn, S.M. A left hemisphere basis for visual mental imagery? *Neuropsychologia* **1985**, *23*, 115–118. [CrossRef]
65. Fulford, J.; Milton, F.; Salas, D.; Smith, A.; Simler, A.; Winlove, C.; Zeman, A. The neural correlates of visual imagery vividness—An fMRI study and literature review. *Cortex* **2018**, *105*, 26–40. [CrossRef] [PubMed]
66. Kosslyn, S.M.; Holtzman, J.D.; Farah, M.J.; Gazzaniga, M.S. A computational analysis of mental image generation: Evidence from functional dissociations in split-brain patients. *J. Exp. Psychol. Gen.* **1985**, *114*, 311–341. [CrossRef]
67. Trojano, L.; Grossi, D. A critical review of mental imagery defects. *Brain Cogn.* **1994**, *24*, 213–243. [CrossRef]
68. Winlove, C.I.; Milton, F.; Ranson, J.; Fulford, J.; MacKisack, M.; Macpherson, F.; Zeman, A. The neural correlates of visual imagery: A co-ordinate-based meta-analysis. *Cortex* **2018**, *105*, 4–25. [CrossRef] [PubMed]
69. De Gelder, B.; Tamietto, M.; Pegna, A.J.; Van den Stock, J. Visual imagery influences brain responses to visual stimulation in bilateral cortical blindness. *Cortex* **2015**, *72*, 15–26. [CrossRef] [PubMed]
70. Gerlach, C.; Klargaard, S.K.; Alnæs, D.; Kolskår, K.K.; Karstoft, J.; Westlye, L.T.; Starrfelt, R. Left hemisphere abnormalities in developmental prosopagnosia when looking at faces but not words. *Brain Commun.* **2019**, *1*, fcz034. [CrossRef]
71. Dinkelacker, V.; Grüter, M.; Klaver, P.; Grüter, T.; Specht, K.; Weis, S.; Kennerknecht, I.; Elger, C.E.; Fernandez, G. Congenital prosopagnosia: Multistage anatomical and functional deficits in face processing circuitry. *J. Neurol.* **2011**, *258*, 770–782. [CrossRef]
72. Dobel, C.; Putsche, C.; Zwitserlood, P.; Junghöfer, M. Early left-hemispheric dysfunction of face processing in congenital prosopagnosia: An MEG study. *PLoS ONE* **2008**, *3*, e2326. [CrossRef]
73. Grüter, T.; Grüter, M.; Bell, V.; Carbon, C.C. Visual mental imagery in congenital prosopagnosia. *Neurosci. Lett.* **2009**, *453*, 135–140. [CrossRef]
74. Barton, J.J.; Cherkasova, M. Face imagery and its relation to perception and covert recognition in prosopagnosia. *Neurology* **2003**, *61*, 220–225. [CrossRef]
75. Rossion, B.; Dricot, L.; Devolder, A.; Bodart, J.M.; Crommelinck, M.; Gelder, B.D.; Zoontjes, R. Hemispheric asymmetries for whole-based and part-based face processing in the human fusiform gyrus. *J. Cogn. Neurosci.* **2000**, *12*, 793–802. [CrossRef]
76. Sack, A.T.; Camprodon, J.A.; Pascual-Leone, A.; Goebel, R. The dynamics of interhemispheric compensatory processes in mental imagery. *Science* **2005**, *308*, 702–704. [CrossRef]

77. Palermo, L.; Nori, R.; Piccardi, L.; Giusberti, F.; Guariglia, C. Environment and object mental images in patients with representational neglect: Two case reports. *J. Int. Neuropsychol. Soc.* **2010**, *16*, 921–932. [CrossRef]
78. Cui, X.; Jeter, C.B.; Yang, D.; Montague, P.R.; Eagleman, D.M. Vividness of mental imagery: Individual variability can be measured objectively. *Vis. Res.* **2007**, *47*, 474–478. [CrossRef]
79. Dijkstra, N.; Bosch, S.E.; van Gerven, M.A. Vividness of visual imagery depends on the neural overlap with perception in visual areas. *J. Neurosci.* **2017**, *37*, 1367–1373. [CrossRef]
80. Nanay, B. Multimodal mental imagery. *Cortex* **2018**, *105*, 125–134. [CrossRef]
81. Grossi, D.; Modafferi, A.; Pelosi, L.; Trojano, L. On the different roles of the cerebral hemispheres in mental imagery: The "o'Clock Test." in two clinical cases. *Brain Cogn.* **1989**, *10*, 18–27. [CrossRef]
82. Paivio, A. Comparisons of mental clocks. *J. Exp. Psychol. Hum. Percept. Perform.* **1978**, *4*, 61–71. [CrossRef]
83. El Haj, M.; Gallouj, K.; Antoine, P. Mental imagery and autobiographical memory in Alzheimer's disease. *Neuropsychology* **2019**, *33*, 609–616. [CrossRef]
84. Bartolomeo, P.; Bachoud-Lévi, A.C.; Chokron, S.; Degos, J.D. Visually- and motor-based knowledge of letters: Evidence from a pure alexic patient. *Neuropsychologia* **2002**, *40*, 1363–1371. [CrossRef]
85. Ogden, J.A. Visual object agnosia, prosopagnosia, achromatopsia, loss of visual imagery, and autobiographical amnesia following recovery from cortical blindness: Case MH. *Neuropsychologia* **1993**, *31*, 571–589. [CrossRef]
86. Greenberg, D.L.; Eacott, M.J.; Brechin, D.; Rubin, D.C. Visual memory loss and autobiographical amnesia: A case study. *Neuropsychologia* **2005**, *43*, 1493–1502. [CrossRef]
87. Pearson, J.; Clifford, C.W.; Tong, F. The functional impact of mental imagery on conscious perception. *Curr. Biol.* **2008**, *18*, 982–986. [CrossRef]
88. Pearson, J. New directions in mental-imagery research. *Curr. Dir. Psychol. Sci.* **2014**, *23*, 178–183. [CrossRef]
89. Boccia, M.; Sulpizio, V.; Palermo, L.; Piccardi, L.; Guariglia, C.; Galati, G. I can see where you would be: Patterns of fMRI activity reveal imagined landmarks. *Neuroimage* **2017**, *144*, 174–182. [CrossRef]
90. Dijkstra, N.; Ambrogioni, L.; van Gerven, M. Neural dynamics of perceptual inference and its reversal during imagery. *BioRxiv* **2019**, 781294. [CrossRef]
91. Lawrence, S.J.; Formisano, E.; Muckli, L.; de Lange, F.P. Laminar fMRI: Applications for cognitive neuroscience. *Neuroimage* **2019**, *197*, 785–791. [CrossRef]
92. Bergmann, J.; Morgan, A.T.; Muckli, L. Two distinct feedback codes in V1 for 'real'and 'imaginary'internal experiences. *BioRxiv* **2019**, 664870. [CrossRef]

Perspective

Insights from a Bibliometric Analysis of Vividness and Its Links with Consciousness and Mental Imagery

Stefanie Haustein [1,2,3], André Vellino [1,*] and Amedeo D'Angiulli [4,5]

1 School of Information Studies, University of Ottawa, Ottawa, ON K1N 6N5, Canada; stefanie.haustein@uottawa.ca
2 Centre Interuniversitaire de Recherche sur la Science et la Technologie (CIRST), Université du Québec à Montréal, Montreal, QC H3C 3P8, Canada
3 Scholarly Communications Lab., Ottawa, ON K1N 6N5, Canada
4 Department of Neuroscience, Carleton University, Ottawa, ON K1S 5B6, Canada; amedeo.dangiulli@carleton.ca
5 NICER Lab., Carleton University, Ottawa, ON K1S 5B6, Canada
* Correspondence: avellino@uottawa.ca

Received: 10 December 2019; Accepted: 7 January 2020; Published: 10 January 2020

Abstract: We performed a bibliometric analysis of the peer-reviewed literature on vividness between 1900 and 2019 indexed by the Web of Science and compared it with the same analysis of publications on consciousness and mental imagery. While we observed a similarity between the citation growth rates for publications about each of these three subjects, our analysis shows that these concepts rarely overlap (co-occur) in the literature, revealing a surprising paucity of research about these concepts taken together. A disciplinary analysis shows that the field of Psychology dominates the topic of vividness, even though the total number of publications containing that term is small and the concept occurs in several other disciplines such as Computer Science and Artificial Intelligence. The present findings suggest that without a coherent unitary framework for the use of vividness in research, important opportunities for advancing the field might be missed. In contrast, we suggest that an evidence-based framework (such as the bibliometric analytic methods as exemplified here) will help to guide research from all disciplines that are concerned with vividness and help to resolve the challenge of epistemic incommensurability amongst published research in multidisciplinary fields.

Keywords: vividness; consciousness; mental imagery; bibliometrics; map of science; term co-occurrence

---

1. Introduction

From the inception of scientific psychology, vividness has been a key psychological construct related to the nature of the human condition and consciousness (see the pioneers such as Galton, Wundt, Hebb). Ubiquitously associated with personality and individual differences of imagination, this concept has had a dubious status and unclear definitions and ever-changing meanings continue into the present. Most recently, a prepotent revival of interest as reflected by almost 400 publications in the last five years (according to the Web of Science) in social, cognitive and neuroscientific psychology poses the challenge of understanding its uses and definitions to create theories that provide a framework for useful knowledge creation.

In many studies, mental images are either treated as conscious entities by definition, or as empirical operations implicit in completing some type of task, such as, for example, the measurement of reaction time in mental rotation [1]. In the latter context, an underlying mental image is assumed, but there is no direct determination of whether it is conscious or not. As summarized by Baars [2], "we have very little firm evidence about the conscious dimension of mental imagery". Vividness of mental images is a construct which may be crucial in achieving this missing bridge in research, as it may correspond to consciousness or aspects of consciousness of images.

There is currently a surge of interest in vividness in the cognitive neuroscience and neuroimaging literature (see [3] for a review). Based on this literature, it seems that a general implicit assumption is that vivid images are conscious, and it is possible that the least vivid images are effectively unconscious or that they become such once a threshold (e.g., the "no image" in the Vividness of Visual Imagery Questionnaire [4]) is reached. Thus, it is still unclear whether the vividness dimension may in fact be a kind of "disguised correlate of consciousness" [2] or if, instead, it might be a supramodal metacognitive dimension not necessarily associated with imagery. However, even from studies using a vividness approach in neuroscience, the conscious dimension of mental imagery is not explicitly or fully tackled head on and the potentially critical contribution of vividness as a mediating property of consciousness is lost. Therefore, as it stands, we seem to be missing pieces in a complex relationship (between consciousness and imagery) which could be captured by a potential mediator (vividness).

To address the pressing question of what we know about the links between vividness, consciousness and imagery in the research done so far, we undertook a bibliometric and terminological analysis of the concept of *vividness* in the context of research in *consciousness* and *mental imagery*, from 1900 to 2019. We performed an analysis of publications indexed in the Web of Science (WoS) containing these terms to both generate a network graph of word co-occurrences and to explore several bibliographic characteristics of these publications over time. Our aim was to create a preliminary overview of the publication landscape of vividness and its links to consciousness and imagery. The motivation was to lay the foundations for further research with the long-term objective of building a unified framework for the cohesive use of the construct of vividness. The rationale is that such a framework could help researchers at least better understand the semantic variations in the current usage of this concept, if not help to formulate a sort of inter- and multi-disciplinary lingua franca.

2. Materials and Methods

The most elementary way to measure of the relevance of an idea or concept in the literature is to calculate the frequency with which peer-reviewed publications mentioning that concept are published. Queries on such terms in the WoS publication metadata of most of the extant literature provided us with these basic publications statistics. The WoS Core Collection used for our analysis includes the Science Citation Index, Social Science Citation Index and the Arts & Humanities Citation Index, which cover 21,177 peer-reviewed journals. At the time of writing, WoS included 74.9 million documents published between 1900 and 2019.

Basic bibliometric indicators, such as the number of publications and co-occurrences, were used to assess the growth of the relevant concepts over time. The analyzed metadata is based on the bibliographic information provided in WoS, which we downloaded from the online platform in September 2019.

Based on the bibliographic information, we constructed concept maps based on term co-occurrences in the title of these papers. We used VOSviewer, a social network analysis software tool specifically designed to extract meaningful network information from bibliographic and bibliometric metadata [5,6]. VOSviewer visualizes network data in the form of network graphs to create effective and efficient visualizations of complex relationship data. For example, one can create co-authorship networks, citation networks and co-occurrence networks. In order to restrict the text to the most meaningful terms, we used VOSviewer to extract noun phrases by applying a linguistic filter based on a part-of-speech tagger [6]. Extracted terms were analyzed and cleaned using the thesaurus function of VOSviewer. We combined acronyms with the occurrences of their long forms (e.g., VVIQ, EMDR, VMIQ, PTSD and their corresponding long form expressions), merged synonyms and quasi-synonyms such as "image vividness" and "imagery vividness" and removed extraneous adjectives in some multi-word terms (e.g., "enhanced vividness", "high vividness") to one significant word (e.g., "vividness"). The final network, after merging synonyms, contained 1065 individual noun phrases. Clustering and network layout in VOSviewer are based on the association strength algorithm, which normalizes co-occurrences of two items by their overall occurrences [6]. Clustering identifies the most frequently co-occurring and thus most similar terms and visualizes this similarity by assigning nodes belonging to the same cluster the same color on the map. Terms that appeared frequently in the title of the same documents, and thus have many connections to other terms, are positioned in the center of the network. Those terms that do not co-appear often with others are located at the periphery. The size of the nodes in the network graph represents the number of occurrences of a term and the thickness of edges demonstrates the number of co-occurrences between two nodes.

3. Results

3.1. Analysis of Title Terms

The numbers of documents published in the last 120 years containing the keywords of interest in the title (TI = ("vividness" or "liveliness")/TI = "mental imag*"/TI = "consciousness") are shown in Figure 1. Overall, 22,909 documents contained consciousness in the title were published, followed by 1381 for mental imag* (which includes "mental image", "mental images", "mental imaging", etc.) and 393 documents on "vividness or liveliness". We can assume that if mentioned in the title, that the articles focus on the respective concepts. When the concepts appear in the abstract or keyword field, they might still be relevant but can be considered less central to the article overall.

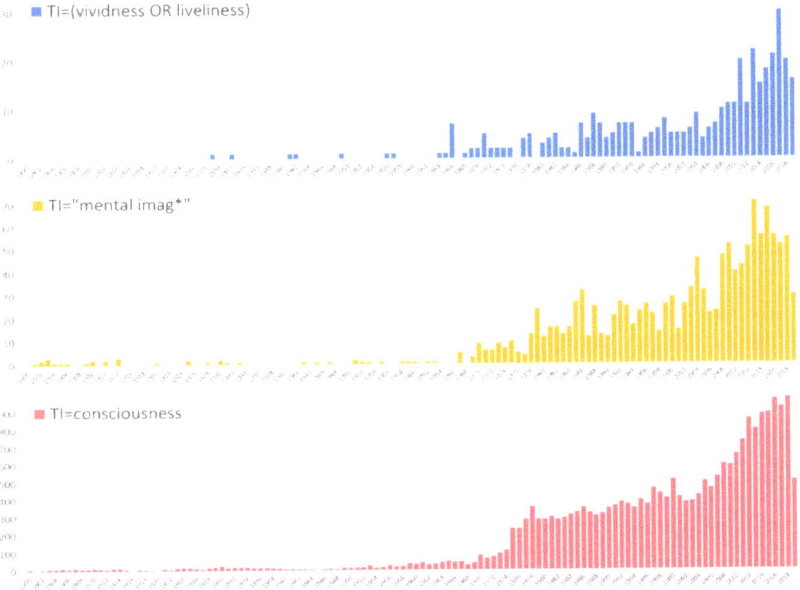

Figure 1. Annual publication rates in Web of Science (WoS) for documents with search terms in the title.

The first publications mentioning "vividness" in the title were both published in the *Journal of Experimental Psychology* in 1929 [7] and 1932 [8]. With the exception of five other early works in in the 1940s and 1950s, the concept only starts to become relevant in the late 1960s. With some declines in 1985 and 1995, the number of publications increased overall, particularly in the mid-2000s. The number of annual publications focusing on vividness peaked in 2017 at 30 documents.

Several features of this figure deserve to be highlighted. The publication rates for each of the search terms from about 1966 onward show similar growth trends. A particular increase can be observed from the mid-2000s onwards. The increase is similar to growth rates in these same fields

(Psychology/Neurosciences) for publications using cognate terms such as "mind" and "mental" (data not shown). We note, however, that there is a time lag between the times at which these concepts reached a critical threshold of relative publication rates. The terms "vividness/liveliness" only reached 10 or more publications per year (roughly 30% of the peak of 30 publications per year in 2017 after 2008) whereas "mental imag*" reached that 30% from its peak per year in 2002 and "consciousness" reached that threshold in 1975.

Looking at the overlap between documents that mention more than one of the concepts in their titles (Figure 2), it becomes apparent that none of the documents address all three concepts together. The size of each circle is proportional to the total number of documents and the size of the intersections is proportional to their co-occurrence in the literature. There are also no documents that mention both vividness (or liveliness) and consciousness in the title. Moreover, the overlap between vividness and mental imag* is low: 32 documents address both concepts in their titles, this corresponds to 8.1% of 393 vividness or 2.3% of mental imaging publications. The overlap between consciousness and mental imag* is even lower at nine documents.

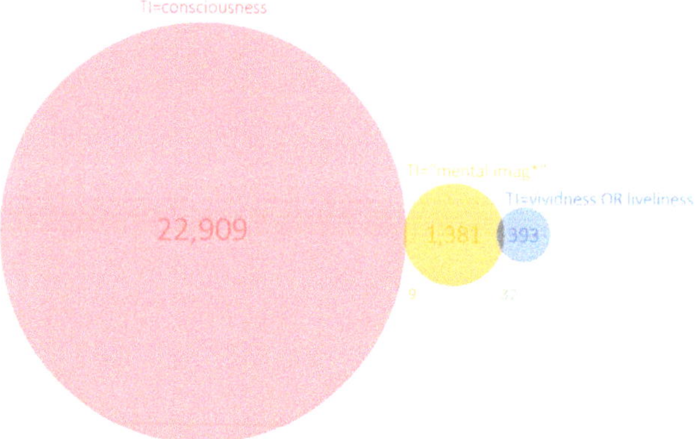

Figure 2. Venn diagram of the co-occurrence of terms in the title field (TI).

On the level of the disciplinary distribution of publications, almost half (48%) of the 393 articles with "vividness" in the title were published in Psychology journals, followed by Neuroscience and Neurology (6%), Psychiatry (5%), Business and Economics (4%) and Literature (3%). On the journal level, articles were much more evenly distributed. As shown in Figure 3, the most common publication venues were the *British Journal of Psychology* (10 documents; 2.5%), *Journal of Behavior Therapy and Experimental Psychiatry* (9; 2.3%), *International Journal of Clinical and Experimental Hypnosis* (7; 1.8%) and *Memory* (6; 1.5%).

The co-occurrence network in Figure 4 shows that in the whole network of relationships there seem to be many connections between the main vividness node and the main nodes for concepts related to imagery, memory and consciousness. However, these connections are scattered and preponderantly weak. The most frequent noun phrases (as displayed by node size) were *vividness, liveliness, effect, imagery, imagery vividness, mental imagery, relationship, VVIQ, visual imagery, memory, individual difference, emotionality, eye movement, memory vividness* and *vividness effect*. The 1065 title terms were grouped into 67 clusters according to their co-occurrence similarity. The largest cluster (red) grouped the terms *memory* (10 document titles), *emotionality* (9), *eye movement* (9) and *memory vividness* (9). It was followed by cluster 2 (green), which contained the terms *vividness* (200), *animated image* (3), *divine power* (3), *naturalism* (3) and *roman theory* (3). Cluster 3 (blue) contained the terms *function* (6), *implication* (6) and *measure* (5). The title terms *individual difference* (10), *moderating role* (3) and *multisensory imagery vividness* (3) were the most frequent noun phrases in cluster 4 (yellow). Cluster 5 contained *impact* (5), *study* (5), *comparison* (3), *fMRI* (3) and *visual imagery vividness* (3), while *liveliness* (44) together with *privacy* (2) was assigned to cluster 6 (turquoise).

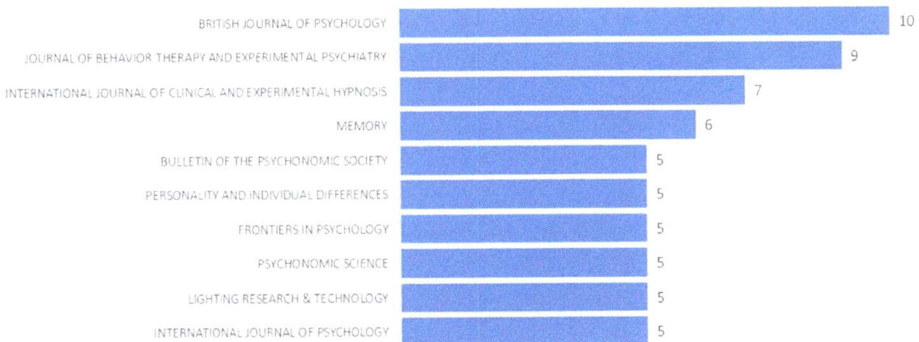

Figure 3. Top 10 journals publishing documents with "vividness" in the title.

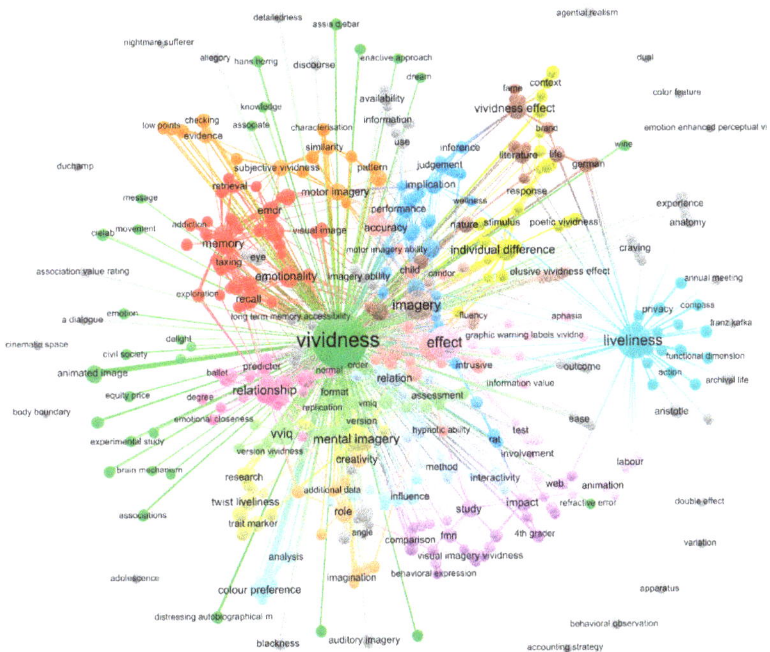

Figure 4. Co-occurrence network of noun phrases based on articles with "vividness" in the title.

3.2. Analysis of Title and Abstract Terms

Expanding the title search to titles, abstracts and keywords (WoS topic search (TS field tag)) includes additional documents which refer to the concepts of vividness, mental imagery and consciousness, but these concepts might not be central to the respective publications. The expansion of the queries increases the number of relevant publications to 2186 for "vividness or liveliness", 5203 for "mental imag*" and 76,920 to "consciousness". We note here that there is proportionally more work published on *vividness* and *mental imagery* than there is on either *consciousness* and *mental imagery* or *consciousness* and *vividness*.

Taken together over time, the number of publications in the intersection of the search terms that co-occur in the title, abstract or keyword fields is shown in the Venn diagram in Figure 5. As little as 15 documents (0.7% of vividness publications) mention all three concepts in the title, abstract or keyword fields. These documents are listed in Table 1 in descending order of citations received. These documents were published in 14 journals between 1999 and 2019 with seven of them published since 2016, which suggests that the combination of the three concepts is gaining traction in recent years. However, the scattering of articles across 14 journals suggests that there is no "natural habitat" for these kinds of publications.

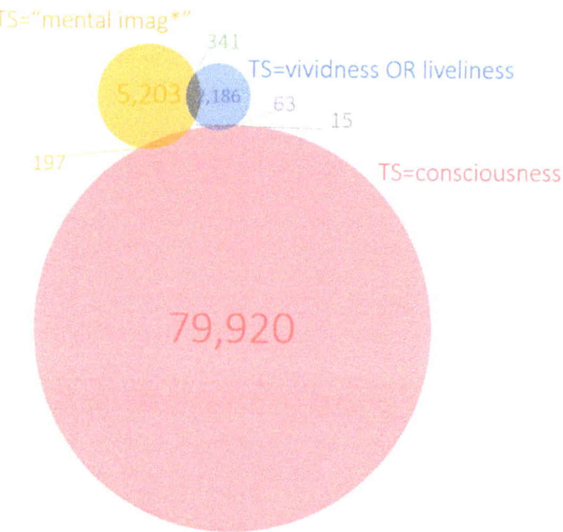

Figure 5. Venn diagram of the co-occurrence of terms in the title, abstract or keyword (TS) fields.

We note also that a subset of these papers argue that the link between these concepts has been neglected over the years. In [3] for example, one of the co-authors of this article argued that a systematic account of mental imagery that integrates its cognitive, affective, neural and phenomenological aspects, including vividness and consciousness is still lacking.

Table 1. Fifteen documents' terms in title, abstract or keywords ordered by number of citations. Matches to search terms in the title are in bold and underscored.

First Author	Title; DOI	Journal	Ref.	WoS Citations
Piolino, P.	Re-experiencing old memories via hippocampus: a PET study of autobiographical memory; 10.1016/j.neuroimage.2004.02.025	*Neuroimage*	[9]	108
Pearson, J.	Evaluating the Mind's Eye: The Metacognition of Visual Imagery; 10.1177/0956797611417134	*Psychological Science*	[10]	59
Marks, D.F.	**Consciousness**, **mental imagery** and action; 10.1348/000712699161639	*British Journal of Psychology*	[11]	59
Lilley, S.A.	Visuospatial working memory interference with recollections of trauma; 10.1348/014466508X398943	*British Journal of Clinical Psychology*	[12]	55
Rademaker, R.L.	Training visual imagery: Improvements of metacognition, but not imagery strength; 10.3389/fpsyg.2012.00224	*Frontiers in Psychology*	[13]	28

Table 1. *Cont.*

First Author	Title; DOI	Journal	Ref.	WoS Citations
Vianna, E.P.M.	Does vivid emotional imagery depend on body signals?; 10.1016/j.ijpsycho.2008.01.013	*International Journal of Psychophysiology*	[14]	8
Huang, M.P.	Vivid visualization in the experience of phobia in virtual environments: Preliminary results; 10.1089/10949310050078742	*Cyberpsychology & Behavior*	[15]	8
Iachini, T.	The experience of virtual reality: are individual differences in **mental imagery** associated with sense of presence?; 10.1007/s10339-018-0897-y	*Cognitive Processing*	[16]	4
Deroy, O.	Lessons of synaesthesia for **consciousness**: Learning from the exception, rather than the general; 10.1016/j.neuropsychologia.2015.08.005	*Neuropsychologia*	[17]	4
Runge, M.S.	Meta-analytic comparison of trial-versus questionnaire-based **vividness** reportability across behavioral, cognitive and neural measurements of imagery; 10.1093/nc/nix006 10.1093/nc/nix006	*Neuroscience of Consciousness*	[3]	3
Santarpia, A.	Evaluating the **vividness** of **mental imagery** in different French samples; 10.1016/j.prps.2007.11.001	*Pratiques Psychologique*	[18]	3
Fazekas, P.	White dreams are made of colours: What studying contentless dreams can teach about the neural basis of dreaming and conscious experiences; 10.1016/j.smrv.2018.10.005	*Sleep Medicine Reviews*	[19]	1
van Schie, C.C.	When I relive a positive me: Vivid autobiographical memories facilitate autonoetic brain activation and enhance mood; 10.1002/hbm.24742	*Human Brain Mapping*	[20]	0
Marks, D.E.	I Am Conscious, Therefore, I Am: Imagery, Affect, Action, and a General Theory of Behavior; 10.3390/brainsci9050107	*Brain Sciences*	[21]	0
Ribeiro, N.	Investigating on the Methodology Effect When Evaluating Lucid Dream; 10.3389/fpsyg.2016.01306	*Frontiers in Psychology*	[22]	0

A tree-map (Figure 6) of the top 15 most frequent disciplines, in which journal articles containing "vividness or liveliness" were published as a topic, shows that the reach of this concept far exceeds the disciplinary boundaries of Psychology. Articles containing "vividness" have been exported to fields such as Computer Science, Business, Engineering and Education. The treemap, where the size of each rectangle represents the number of articles, is based on WoS categories, which are assigned at the journal level.

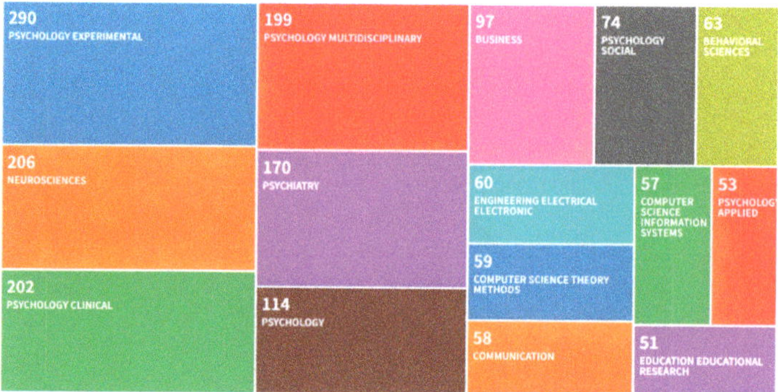

Figure 6. Top 15 disciplines of publications containing the term "vividness" in the title, abstract or keyword fields.

4. Discussion

From our results, it seems evident that a mapping of terms, concepts and constructs linked with vividness through bibliometric analysis provides valuable insights. As a researcher in the fields of Psychology or Neuroscience would expect, vividness appears linked with the subfields of imagery, memory, and clinical practice. However, it also appears that the set vividness is related to many different scattered aspects related to the set imagery/memory, and consciousness. This is evident from Figure 4. Furthermore, confirming the very motivation of our analysis, there are actually few *direct* explicit intersections between vividness, imagery/memory and consciousness, as reflected by the low number of journal articles containing more than one of the three terms (Figure 3). The historical trends of publications in Figure 1 suggest that there might be similar patterns of growth in publication, which could be explained by the common repressive influence of behaviorism on the three constructs as they represent three main examples of "mentalism" [23]. This may be evidence for why the links between these variables might have been expressed only indirectly. Whatever the cause, an implication of the scarcity of intersections between the three constructs in current research is suggestive of the challenge in integrating knowledge about them in an effective manner in the current scientific paradigm.

These findings could have a few possible interpretations. However, we argue that a plausible reason that there are many scattered links between vividness, consciousness and imagery, which are yet not as direct as one might expect, rests in the status of vividness itself as a construct. It is possible to interpret the pattern of results as showing that although vividness is at the center of so much research, it indeed is the least established concept, lagging far behind consciousness and imagery as central concepts in Psychology and Neuroscience. A plausible and parsimonious explanation for this is the polysemy of the term "vividness". Besides the ordinary sense of "vivid" associated with clarity, intensity of hue and chromatic purity, this term is also used as synonym for relevance or salience, emotional expressivity or intensity of emotional content, strength of memory, richness of imagination or detail or meaning, and finally the ease with which memories or mental images are recalled.

The major complicating factors seem to be the surprising variety of meanings of the term vividness and how it is used or theorized. Some authors do not mention imagery or consciousness at all when using the term vividness but associate it with various forms of memory such as prospective, episodic and autobiographical memory, or to aliased processes not literally called imagery (e.g., imaginings, visualizations, simulations). Similarly, replacement constructs for vividness have been offered, for example, in terms of strength of imagery or semantic long-term memory contents, such as the general memory dimensions mentioned earlier or more specifically sensory, autobiographical or episodic memories. In other cases, vividness is replaced by synonyms such as "liveliness" (which we captured in our analysis) or the even vaguer "richness" and used in a way that is purely narrative and disjoined from previous scientific literature.

In future bibliometric research of this kind, it may be useful to expand the above analyses to include more articles without the targeted keywords. Our initial analysis made a reasonable attempt at identifying synonyms for the targeted keywords, but it was not intended to be exhaustive (for instance, some previous work used "clarity" instead of "vividness" or "liveliness"). The analyses could also go beyond the keywords in title and abstract, for instance, by including established experimental paradigms and measures for these three concepts which fall in the purview of the proposed background framework of reference.

Consequently, current research practice related to vividness resembles a Tower of Babel where researchers from different traditions talk over each other without sharing a common language, understanding or knowledge. One possible reason is that while the construct is very salient, its use might have far exceeded the disciplinary boundaries of Psychology and Neuroscience. For instance, it has been exported in fields such as Computer Science, AI and consciousness research, while simultaneously having a rich historical background of usage in many languages and non-scientific cultures (for example in poetry and literature) [24].

If the reasoning above is plausible, there is much to be gained (by researchers in the field) to consolidate the construct of vividness by looking at the meaning of the different occurrences in the various relationships that one can find in the overlapping literatures. Our analysis could be a starting point for this long-term objective. For example, it may be possible to refine the mapping created in Figure 5 to extract a core and periphery of what a normative meaning of vividness could be for multidisciplinary research in consciousness and imagery.

The product of this analysis could be a framework that could be empirically tested and therefore help researchers stir research into productive, useful and plausible scientific directions (i.e., avoiding seemingly contradictory situations of many relational associations which remain like isolated islands and do not reach full integration in a coherent wider network of theoretical statements, models and constructs). Case in point, it seems that if things remain as they are, statements about vividness cannot be easily compared empirically (i.e., they are epistemically incommensurable [25]) from one paper to the next. Yet, contrary to what one may infer from this fragmentation of knowledge, our analysis provides evidence that the construct of vividness does have some explicative power or efficacy in terms of unifying theory.

Thus, it is possible that unless we try to achieve a coherent unitary framework for the use of vividness in research, important opportunities for advancing the field might be missed. On the contrary, an evidence-based framework will help to resolve semantic ambiguities and guide research from all disciplines that are concerned with vividness.

The need for such a framework is exemplified by the limitations of the current bibliometric analysis itself. Our searches for term-occurrence in the WoS does not, in fact, guarantee a maximum amount of "recall" for the topical aboutness of publications. In other words, it is possible to write about a subject ("vividness") without ever mentioning the exact term. Thus, in our study three peer reviewed journals, *Journal of Mental Imagery*, *Imagination Cognition and Personality* and *Memory and Cognition*, are not on the list or are not the main contributors in the list of most relevant journals but do in fact contain articles that are related to vividness even if the exact word does not appear in the metadata of the articles. It is clear that to overcome this limitation in future bibliometric analyses it is essential to have a solid conceptual framework of reference to make valid semantic attributions and inferences.

We conclude that bibliometric analysis, as demonstrated by this preliminary work and further strengthened by semantically enhanced search, could be an invaluable tool for showing future research the best way ahead.

Author Contributions: Conceptualization, S.H., A.V. and A.D.; methodology, S.H. and A.V.; software, S.H.; validation, S.H., A.V. and A.D.; formal analysis, S.H.; investigation, S.H., A.V. and A.D.; resources, S.H., A.V. and A.D.; data curation, S.H., A.V. and A.D.; writing—original draft preparation, A.V and A.D; writing—review and editing, S.H., A.V. and A.D.; visualization, S.H. All authors have read and agreed to the published version of the manuscript.

Funding: This research received no external funding.

Acknowledgments: We would like to thank Leo Holton (McGill, Cognitive Science) and Evan Sterling (University of Ottawa Library) for their initial contributions to this research.

Conflicts of Interest: The authors declare no conflict of interest.

References

1. Shepard, R.N.; Metzler, J. Mental Rotation of Three-Dimensional Objects. *Science* **1971**, *171*, 701–703. [CrossRef]
2. Baars, B.J. Understanding subjectivity: Global workspace theory and the resurrection of the observing self. *J. Conscious Stud.* **1996**, *3*, 211–216.
3. Runge, M.S.; Cheung, M.W.-L.; D'Angiulli, A. Meta-analytic comparison of trial-versus questionnaire-based vividness reportability across behavioral, cognitive and neural measurements of imagery. *Neurosci. Conscious.* **2017**, *2017*, nix006. [CrossRef]
4. Marks, D.F. Visual imagery differences in the recall of pictures. *Br. J. Psychol.* **1973**, *64*, 17–24. [CrossRef]
5. Van Eck, N.J.; Waltman, L. VOSviewer Manual. Available online: https://www.vosviewer.com/documentation/Manual_VOSviewer_1.6.13.pdf (accessed on 16 September 2019).
6. Van Eck, N.J.; Waltman, L. Software survey: VOSviewer, a computer program for bibliometric mapping. *Scientometrics* **2010**, *84*, 523–538. [CrossRef] [PubMed]
7. Jersild, A. Primacy, recency, frequency and vividness. *J. Exp. Psychol.* **1929**, *12*, 58–70. [CrossRef]
8. Van Buskirk, W.L. An experimental study of vividness in learning and retention. *J. Exp. Psychol.* **1932**, *15*, 563–573. [CrossRef]
9. Piolino, P.; Giffard-Quillon, G.; Desgranges, B.; Chételat, G.; Baron, J.C.; Eustache, F. Re-experiencing old memories via hippocampus: A PET study of autobiographical memory. *Neuroimage* **2004**, *22*, 1371–1383. [CrossRef]
10. Pearson, J.; Rademaker, R.L.; Tong, F. Evaluating the mind's eye: The metacognition of visual imagery. *Psychol. Sci.* **2011**, *22*, 1535–1542. [CrossRef]
11. Marks, D.F. Consciousness, mental imagery and action. *Br. J. Psychol.* **1999**, *90*, 567–585. [CrossRef]
12. Lilley, S.A.; Andrade, J.; Turpin, G.; Sabin-Farrell, R.; Holmes, E.A. Visuospatial working memory interference with recollections of trauma. *Br. J. Clin. Psychol.* **2009**, *48*, 309–321. [CrossRef] [PubMed]
13. Rademaker, R.L.; Pearson, J. Training visual imagery: Improvements of metacognition, but not imagery strength. *Front. Psychol.* **2012**, *10*, 224. [CrossRef] [PubMed]
14. Vianna, E.P.; Naqvi, N.; Bechara, A.; Tranel, D. Does vivid emotional imagery depend on body signals? *Int. J. Psychophys.* **2009**, *72*, 46–50. [CrossRef] [PubMed]
15. Huang, M.P.; Himle, J.; Alessi, N.E. Vivid visualization in the experience of phobia in virtual environments: Preliminary results. *Cyberpsychol. Behav.* **2000**, *3*, 315–320. [CrossRef]
16. Iachini, T.; Maffei, L.; Masullo, M.; Senese, V.P.; Rapuano, M.; Pascale, A.; Sorrentino, F.; Ruggiero, G. The experience of virtual reality: Are individual differences in mental imagery associated with sense of presence? *Cognit. Process.* **2019**, *20*, 291–298. [CrossRef]
17. Deroy, O.; Spence, C. Lessons of synaesthesia for consciousness: Learning from the exception, rather than the general. *Neuropsychologia* **2016**, *88*, 49–57. [CrossRef]
18. Santarpia, A.; Blanchet, A.; Poinsot, R.; Lambert, J.F.; Mininni, G.; Thizon-Vidal, S. Evaluating the vividness of mental imagery in different French samples. *Pratiques Psychologiques* **2008**, *14*, 421–441. [CrossRef]
19. Fazekas, P.; Nemeth, G.; Overgaard, M. White dreams are made of colours: What studying contentless dreams can teach about the neural basis of dreaming and conscious experiences. *Sleep Med. Rev.* **2019**, *43*, 84–91. [CrossRef]
20. Van Schie, C.C.; Chiu, C.D.; Rombouts, S.A.; Heiser, W.J.; Elzinga, B.M. When I relive a positive me: Vivid autobiographical memories facilitate autonoetic brain activation and enhance mood. *Hum. Brain Map.* **2019**, *40*, 4859–4871. [CrossRef]
21. Marks, D.F. I Am Conscious, Therefore, I Am: Imagery, Affect, Action, and a General Theory of Behavior. *Brain Sci.* **2019**, *9*, 107. [CrossRef]
22. Ribeiro, N.; Gounden, Y.; Quaglino, V. Investigating on the methodology effect when evaluating lucid dream. *Front. Psychol.* **2016**, *7*, 1306. [CrossRef] [PubMed]

23. Paivio, A. Neomentalism. *Can. J. Psychol.* **1975**, *29*, 263–291. [CrossRef]
24. Parker, P. *Vividness—Webster's Timeline History 353 BC–2007*; ICON Group International Inc.: San Diego, CA, USA, 2010.
25. Feyerabend, P. Explanation, Reduction and Empiricism. In *Scientific Explanation, Space, and Time*; Minnesota Studies in the Philosophy of Science; Feigl, H., Maxwell, G., Eds.; University of Minneapolis Press: Minneapolis, MN, USA, 1962; Volume III, pp. 28–97.

MDPI
St. Alban-Anlage 66
4052 Basel
Switzerland
Tel. +41 61 683 77 34
Fax +41 61 302 89 18
www.mdpi.com

Brain Sciences Editorial Office
E-mail: brainsci@mdpi.com
www.mdpi.com/journal/brainsci

www.ingramcontent.com/pod-product-compliance
Lightning Source LLC
LaVergne TN
LVHW070839170726
843515LV00005B/525

* 9 7 8 3 0 3 6 5 0 4 1 2 4 *